APOLLO'S LEGACY

ROGER D. LAUNIUS

APOLLO'S LEGACY

PERSPECTIVES ON THE MOON LANDINGS

Smithsonian
Books

WASHINGTON, DC

This book may be purchased for educational, business, or sales promotional use. For information, please write Special Markets Department, Smithsonian Books, P.O. Box 37012, MRC 513, Washington, DC 20013

Published by Smithsonian Books

Director: Carolyn Gleason

Creative Director: Jody Billert

Senior Editor: Christina Wiginton

Editor: Laura Harger

Editorial Assistant: Jaime Schwender

Edited by Gregory McNamee

Typeset and indexed by Scribe Inc.

Library of Congress Cataloging-in-Publication Data
Names: Launius, Roger D., author.
Title: Apollo's legacy : perspectives on the moon landings / Roger D. Launius.
Description: Washington, DC : Smithsonian Books, 2019. |
Includes bibliographical references and index.
Identifiers: LCCN 2018032602 (print) | LCCN 2018037503 (ebook) |
ISBN 9781588346520 (eBook) | ISBN 9781588346490 (hardcover)
Subjects: LCSH: Project Apollo (U.S.)—History. | Space flight to
the moon—History. | Space race—United States—History. |
Outer space—Exploration—United States—History.
Classification: LCC TL789.8.U6 (ebook) | LCC TL789.8.U6 L27 2019 (print) |
DDC 629.45/4—dc23
LC record available at https://lccn.loc.gov/2018032602

Manufactured in the United States of America

23 22 21 20 19   5 4 3 2 1

To the memory of my favorite astronauts:
Pete Conrad (1930–1999) and Sally Ride (1951–2012)

Buzz Aldrin: There it is, it's coming up!
Michael Collins: What?
Aldrin: The Earth. See it?
Collins: Yes. Beautiful.
Neil Armstrong: The *Eagle* has wings.

APOLLO II CREW IN CONVERSATION DURING
MOON APPROACH, JULY 20, 1969

CONTENTS

PROLOGUE

July 20, 1969

The astronauts who first landed on the Moon half a century ago carried with them the hopes and wishes of all whom they had left behind on Earth, as well as uncertainty about what they would experience on the lunar surface. Setting foot on another world, they knew, would be the climax of humanity's greatest adventure to date. Yet they betrayed none of this as they went about their work. They had trained for this mission. They had drilled for every contingency. In every way possible, they had worked to minimize risk. But risk there was, and they understood it, accepted it, and even welcomed it.

The lunar module (LM), code-named *Eagle*, that brought astronauts Neil Armstrong and Buzz Aldrin to the surface of the Moon that day was one of the weirdest-looking contraptions ever to invade the sky. A completely nonaerodynamic multisided box with the tensile strength of aluminum foil, floating with its gangly legs jutting out, it possessed neither symmetry nor grace. Mike Collins, piloting *Columbia*, the separate Apollo command module, above the Moon, referred obliquely to its awkward appearance as he joked to Aldrin and Armstrong, "I think you've got a fine-looking flying machine there, *Eagle*, despite the fact you're upside down."

In the first of two maneuvers they undertook to reach the surface, Armstrong and Aldrin, strapped into *Eagle* standing up, descended to 50,000 feet. Next, if all went according to plan, *Eagle* would glide along a 12-minute computer-controlled descent to a point at which Armstrong would take over for a manual landing. But things didn't go according to plan.

As Armstrong and Aldrin descended to 6,000 feet above the lunar surface, a yellow caution light blinked. Then another came on, and another, and another. Even so, Steve Bales, the flight controller at NASA Mission Control in Houston who was responsible for the LM's computers, told the astronauts to proceed. It was just a "1202" program alarm, he reassured them. Collins could not remember the meaning of the 1202 alarm and looked it up as the lander continued down to the lunar surface, later recalling: "My checklist says 1202 is an 'executive overflow,' meaning simply that the computer has been called upon to do too many things at once and is forced to postpone some of them. A little farther along, at just three thousand feet above the surface, the computer flashes 1201, another overflow condition, and again the ground is superquick to respond with reassurances." Everyone has seen such problems with computers in the 21st century as well: the machine freezes up, going to a blue screen or a spinning beach ball. Such situations are irksome, but they are rarely perilous. If the computer helping to control the LM failed, piloting it would be more difficult—and potentially dangerous for the astronauts—but that it turned out to be mostly an annoyance was thanks to the skill and cool-headedness of Armstrong and Aldrin.

The two descending astronauts spotted a landing site inside a large crater, but as they approached, the alarms continued, and they encountered a boulder field. Armstrong piloted *Eagle* away from the danger as Aldrin called out their steadily reduced altitude and range:

Altitude-velocity light. 3½ down, 220 feet, 13 forward. . . .
Coming down nicely. 200 feet, 4½ down. 5½ down. . . . That's
good. 120 feet. . . . There's looking good. Down a half,
6 forward. . . . Lights on. . . . Good. 40 feet, down 2½. Kicking
up some dust. 30 feet, 2½ down. Faint shadow. 4 forward. . . .
Drifting to the right a little. Okay.

Then Aldrin uttered the first words ever spoken on the Moon: "Contact light." By that, he meant that the sensor on one of *Eagle*'s legs had encountered the surface and triggered a light inside the lander.

Armstrong called Mission Control at 3:18 Central Daylight Time: "Houston, Tranquillity Base here, the *Eagle* has landed."

Astronaut Charlie Duke, back at Mission Control, breathed a sigh of relief. "Roger, Tranquillity," he said in his thick southern drawl. "We copy you on the ground. You've got a bunch of guys about to turn blue. We're breathing again. Thanks a lot."

The drama unfolded from there, Armstrong and Aldrin performing their scripted duties as the LM settled further onto the lunar surface. A few hours later, tasks complete, they opened the hatch, and Armstrong squeezed himself out and began to descend the ladder. "That's one small step for [a] man, one giant leap for mankind," Armstrong said as he moved one bulky white boot off the *Eagle*'s footpad. There would be debate about his precise phrasing: no one heard the "a" in "a man" at the time, and Armstrong later explained that in the excitement of the moment, he'd misstated what he had intended to say, and he welcomed the bracketed insertion whenever the phrase appeared thereafter. Regardless of their exact form, these few words were America's announcement, through Armstrong's voice, that it had won the competition that had consumed NASA for years. The two most powerful nations on Earth, the United States and the Soviet Union, had sought to extend their terrestrial control to the Moon, and the United States had used the Apollo program as a key diplomatic tool in the Cold War.

Both astronauts ignored the political dimensions of the moment. They had a lengthy list of duties to perform during their relatively short two-hour moonwalk. As Armstrong would later say, "There were a lot of things to do, and we had a hard time getting them finished. . . . The primary difficulty was just far too little time to do the variety of things we would have liked. We had the problem of the five-year-old boy in a candy store."

They set out scientific experiments and collected rock and soil samples, but they also had two important ceremonial tasks. The first was the unfurling of the American flag on the lunar surface. An American flag had been displayed somewhere in almost every phase of the Apollo 11 flight, yet planting that flag into the dust of the Moon, saluting it, taking pictures of it, and then *not* claiming the Moon for the United States (as had long been the norm when new lands were encountered in the course of imperial expansion) was a deeply symbolic gesture. It proved harder than first thought to unfurl the flag on the lunar surface. A telescoping horizontal rod at the top of the pole would hold the flag out in the airless, windless lunar environment for all to see, but the astronauts rotated the pole and bent the rod as they tried to deploy it, which resulted in the flag's looking a bit like it was blowing in a wind. (Moon-landing deniers have charged that this proved their nonsensical assertion that the landing was filmed on a soundstage and the flag was indeed blowing in the wind. The flag's rippled look did impress other astronauts, who intentionally bent the rods on their flagpoles just a bit to get the same effect.) The Apollo 11

Buzz Aldrin at the flag, an iconic image from Apollo 11. This image circled the globe immediately after its release in July 1969 and has been used for all manner of purposes since that time. The flag in this image proved a powerful trope of American exceptionalism, but the photo also has been used by Moon-landing deniers as "evidence" that the landing was filmed on Earth. The flag appears to be waving in a breeze—and, as we all know, there is no breeze on the Moon—but the simple truth is that as the astronauts were planting the flagpole, they rotated it back and forth to better penetrate the lunar soil (anyone who's set a blunt tent post knows how this works) and bent the telescoping rod. The flag waved as a result—no breeze required. (NASA image no. AS11-40-5875)

astronauts further underlined their peaceful message of exploration by unveiling a plaque on the ladder of the LM that announced:

Here men from the planet Earth
first set foot upon the Moon
July 1969, A.D.
We came in peace for all mankind.

The three-man Apollo 11 crew—Armstrong, Aldrin, and Collins—proved themselves excellent protagonists in this premier lunar drama. Armstrong, in his role as mission commander, logically had become the first to set foot on the Moon. The quintessential flyer, he betrayed no bravado as he went about his tasks in the same businesslike manner with which he approached everything else. Never one to seek the spotlight, Armstrong took his greatest pride in his naval service during the Korean War, when he had flown combat missions from the USS *Essex* against targets immortalized in the James Michener novel *The Bridges at Toko-Ri*, and in his work as a research pilot, when he had flown high-performance aircraft such as the X-15. He was always more comfortable with a small group of friends than he was in the limelight before millions. He could have done whatever he wished after his completion of the Apollo 11 Moon landing and the attendant storm of fame that greeted the astronauts' return to Earth, but Armstrong, modestly, chose to teach aerospace engineering at the University of Cincinnati.

As quiet and circumspect as Armstrong was, Aldrin was just that extroverted. While Armstrong eschewed the spotlight, it attracted Aldrin like a moth to the flame. He never met a camera he did not want to stand in front of, a celebrity golf tournament or *Dancing with the Stars* television competition in which he did not want to indulge his ego. At the same time, no astronaut played a larger role in helping America reach the Moon in 1969. Aldrin held a PhD in astronautics from the Massachusetts Institute of Technology, and his knowledge served the mission well in ensuring the correct maneuvering, rendezvous, and docking of two spacecraft in orbit. His work on extravehicular activity (EVA), or spacewalking, made it possible for astronauts to depart their spacecraft and undertake pioneering work in the vacuum.

While Armstrong and Aldrin ventured onto the Moon's surface, Collins remained in the *Columbia* command module, orbiting the lifeless gray orb. He had the loneliest job in the universe. He recalled later that this experience remained fundamental to his perceptions ever after, eloquently describing his realization of the fragility of Earth as he viewed it from his solitary perch:

From space there is no hint of ruggedness to it [Earth]; smooth as a billiard ball, it seems delicately poised on its circular journey around the Sun, and above all it seems fragile. . . . Is the sea water clean enough to pour over your head, or is there a glaze of oil on its surface? . . . Is the riverbank a delight or an obscenity? The difference between a blue-and-white planet and a black-and-brown one is delicate indeed.

Collins's sensitivity and self-reflection stand in stark contrast to both the modest practicality of Armstrong and the boisterousness of Aldrin. Still, the three men made a good team, conducting humanity's first mission to land on the Moon with efficiency and effectiveness. They were only the point of a spear made up of thousands of NASA engineers and scientists, managers and administrators, clerks and technicians who underpinned the whole Apollo program. During the early 1960s, as the program got underway, the space agency's civil service roll had grown from 10,000 to 36,000 people, and because NASA's leaders made an early decision that they would rely upon outside researchers and technicians to complete Apollo, contractor employees working on the program increased more than tenfold, from 36,500 in 1960 to 376,700 in 1965. Private industry, research institutions, and universities—not government employees per se—provided most of the personnel working on Apollo.

The devotion of this variegated workforce to making Apollo a success is legendary. Those who were part of the program have all told similar stories about it, albeit stating them in different ways and emphasizing different players, that highlight their sense of shared purpose and their belief that they all were active contributors to Apollo's achievements. In one such recitation, which may be apocryphal, a journalist asked an assortment of workers about their jobs. Depending on an interviewee's trade, workers would respond with "I'm a cook," "I'm a salesman," or the like. When the journalist asked a janitor at NASA about *his* job, though, the man replied, "I'm helping to put a man on the Moon." This anecdote highlights two very real traits among those engaged in helping the astronauts reach the lunar surface: their identification with the mission of the organization they worked for, and their recognition of its importance.

Accomplishing this mission had been no easy task. On May 25, 1961, President John F. Kennedy had committed the United States, "before this decade is out, [to] landing a man on the moon and returning him safely to the earth." His call did not herald a bold initiative conducted in the cause of pure science, however. Instead, it was meant to counter a threat in the ominous Cold War clouds then hovering over the United States and the Soviet Union as each nation sought to demonstrate to a divided world its technological mastery and thereby gain the upper hand in the East-West rivalry. As Kennedy made clear in his announcement of the Apollo endeavor, the Moon landing was meant to exhibit, in spectacular fashion, American technological prowess: "No single space project in this period will be more impressive to mankind, or more important for the long-range exploration of space; and none will be so difficult or expensive to accomplish," he said. NASA went to work immediately to achieve this end, and through all manner of exertion, trial, and success, by July 1969 it had come to the point where the first crew could land on the Moon's surface.

While the NASA workforce hunkered down, pursuing the quest for the Moon, the world around it was changing. The series of American confrontations with the Soviet Union in 1961 and 1962—including the failed Bay of Pigs invasion and the Cuban Missile Crisis—was giving way to a looser rapprochement. By the fall of 1963, the international situation had eased so much that Kennedy proposed making the Moon-landing effort a joint program with the Soviet Union in a September speech to the United Nations, declaring:

> There is room for new cooperation, for further joint efforts
> in the regulation and exploration of space. I include among these
> possibilities a joint expedition to the moon. . . . Why, therefore,
> should man's first flight to the moon be a matter of national
> competition? Why should the United States and the Soviet Union,
> in preparing for such expeditions, become involved in immense
> duplications of research, construction, and expenditure? Surely we
> should explore whether the scientists and astronauts of our two
> countries—indeed of all the world—cannot work together in the
> conquest of space, sending some day in this decade to the moon
> not the representatives of a single nation, but the representatives
> of all of our countries.

It is possible that only Kennedy's assassination on November 22 prevented a joint US-USSR landing from taking place. At the same time, the buttoned-down ethos of the early 1960s—the era of the so-called organization man, the "man in the gray flannel suit," and advertising's "mad men" of Madison Avenue—was giving way to the unrest and counterculture that would define the later 1960s. The civil rights movement was reaching a crescendo, and the quest for egalitarianism across American society, rampant antiestablishment sentiment, and the rise of antiwar activism combined to force social change. As President Lyndon B. Johnson remarked in a speech at the University of Michigan in May 1964:

> For better or for worse, your generation has been appointed by history to deal with those problems and to lead America toward a new age. You have the chance never before afforded to any people in any age. You can help build a society where the demands of morality, and the needs of the spirit, can be realized in the life of the Nation. So, will you join in the battle to give every citizen the full equality which God enjoins and the law requires, whatever his belief, or race, or the color of his skin?

In the same speech, Johnson rolled out a broad domestic agenda aimed at creating the "Great Society," meaning a more equitable one. His plans achieved early success only to be sidetracked by the Vietnam War and civil unrest.

The years 1968 through 1970 were among the most tumultuous in American history. Early 1968 saw the capture of the American surveillance ship USS *Pueblo* by North Korea, which led to an 11-month-long hostage crisis, and the Tet Offensive, in which the North Vietnamese and Viet Cong attacked Saigon and more than 100 other South Vietnamese cities. These events were only the beginning of a dire year. The April assassination of Martin Luther King Jr. by a white supremacist led to riots in more than a dozen major US cities. Candidate Robert M. Kennedy died in early June in a hail of assassin's bullets just after winning the California and South Dakota Democratic primaries. In August, violence at the Democratic National Convention in Chicago further exemplified the troubles of the time, as did the October protest of two African American medalists at the Olympic Games who raised their fists in silent protest against the injustice they saw around them.

In the midst of this national toil and trouble, NASA kept working toward the Moon landing. A lunar orbit on December 24, 1968, preceded it, in which Apollo 8 astronauts Frank Borman, James A. Lovell Jr., and William A. Anders became the first human beings to venture beyond low Earth orbit and visit the outskirts of another world. Apollo 8 had been planned as a mission to test hardware, but senior engineer George M. Low, of NASA's Manned Spacecraft Center in Houston, pressed for approval to make it a circumlunar flight. NASA Associate Administrator for Manned Space Flight George E. Mueller argued in a November 1968 memorandum that a circumlunar flight would "represent a significant new international achievement in space . . . [and] provide a significant boost to the morale of the entire Apollo program, and an impetus which must, inevitably enhance our probability of successful lunar landing in 1969." The results of such a mission could prove important, they reasoned: Technical and scientific knowledge could be gained, and the flight would serve as a public demonstration of what the United States was poised to achieve.

When Apollo 8 arrived over the Moon that Christmas Eve, the astronauts saw a gray and battered wasteland below them and, in the distance, a tiny, lovely, and fragile "blue marble" hanging in the blackness of space: Earth. With this vision came the realization that humankind's home is the only place we can live. The crew's broadcast back to Earth helped to bring together humanity, if only for a brief time, as the astronauts read from the first part of the Bible—"God created the heavens and the Earth, and the Earth was without form and void"—before sending Christmas greetings to humanity. On Christmas Day, Apollo 8 headed back to Earth. There its commander, Borman, received a telegram, one of many from fans around the world, that connected his nation's unrest to the mission itself: "Thanks, you saved 1968."

The climactic chapter of the Apollo saga, the first Moon landing, was a must-see event the world over. No one had seen anything like it before, and many would later say the achievement eclipsed any other event of global note in their lifetimes. Some compared Neil Armstrong's reaching the Moon to Christopher Columbus's reaching the Americas across a partially uncharted ocean: both were vanguards of sustained exploration and settlement. NASA's rocket developer Wernher von Braun compared the landing to the moment when the first creature left the sea for dry land. President Richard M. Nixon went even further: "This is the greatest week since the beginning of the world, the creation."

All these were overstatements, but virtually everyone embraced the flight of Apollo 11 as a shared success for the planet. NASA estimated that because of nearly worldwide radio and television coverage, more than half the population of Earth was aware of the events of Apollo 11 in near real time. Police reports noted that streets in many cities were eerily quiet during the moonwalk because residents were inside, watching television coverage of the landing. One seven-year-old boy from San Juan, Puerto Rico, said, "I kept racing between the TV and the balcony and looking at the Moon to see if I could see them on the Moon." His experience was typical.

Even in Moscow, a sense of accomplishment briefly reigned. Most Soviets viewed the American astronauts' reaching the Moon as being less a defeat in the space race than it was a triumph for the whole of human-kind. The Soviet newspaper *Pravda* published a front-page story on Apollo 11 within hours. Soviet television also replayed the first Armstrong and Aldrin moonwalk multiple times, but it did not show the live broadcast.

The world felt better about itself, if only for a few days. Apollo 11 represented a major achievement—but the overall Apollo program would end abruptly just three years later, in 1972, after six successful landings. As federal budget cuts loomed and the political context that first gave rise to Apollo shifted, the public seemed to grow weary of the adventure. Some wondered what there might be left to achieve in space exploration, now that Americans had touched down on the Moon. Others wondered whether Armstrong should be seen less as a modern Christopher Columbus and more as a modern Leif Erickson, whose voyages to America were still-born, a dead end in European exploration of new lands.

Such explanations still feel incomplete as we look back over that past five decades to that transformative night in July 1969. While the major contours of the American sprint to the Moon during the 1960s have been told and retold, and Project Apollo itself—the sites where it took place, the people who participated in it, and our memories of it—has been analyzed, critiqued, celebrated, and castigated, depending on one's perspective, many questions remain. What has been the long-term significance

One of the iconic images from the Apollo program, this photograph shows Buzz Aldrin on the lunar surface during the July 20, 1969, Apollo 11 mission. It has been reproduced in many forms around the world. Seen in the foreground is the leg of the lunar module *Eagle*. Astronaut Neil A. Armstrong, commander of the mission, took this photo of Aldrin's moonwalk; his own image is reflected in the visor of Aldrin's spacesuit. (NASA image no. AS11-40-5903)

of the Apollo program now that half a century has passed? How might we interpret the Apollo adventure in the 21st century, in our postmodern world far removed from that of the late 1960s and early 1970s? What do the Moon landings mean to people of differing cultural, generational, economic, social, and ethnic backgrounds? What role did Apollo play at the time—and afterward—in helping to define modern American society, politics, and self-perception? What is it about the Apollo program that has captured and continues to inform the imagination of the American people? Finally, which elements of Apollo retain their saliency as the program recedes into history? This book endeavors to explore these and many other questions about the legacy of Apollo.

In the case of the Moon landings, memory operates in many powerful ways. Most remembrances of the landings have been dominated by a celebratory perspective, one emphasizing the uniqueness that is America as it overcomes challenge and adversity. Triumphalism, as we might term this viewpoint, emphasizes the qualitative difference of the United States from other nations and its distinctive character, forged in the country's revolutionary origins but honed by its focus on liberty, egalitarianism, individualism, and laissez-faire capitalism. The successful Apollo program can be viewed as a representation of American history writ large, and, indeed, this has become the standard way to remember it. This way of recalling Apollo focuses upon an initial shock to the system: a challenge from a powerful Soviet Union, in the form of the Sputnik launch a dozen years earlier, was threatening to swamp American capability, but the United States rose to the test and demonstrated its own power. Ultimately, in this tale, America is proved both justified and triumphant through the achievements of its Apollo program, and thus the dominant American perspective on the past is sustained.

Three counter-stories stand alongside the triumphalist one, however, and they emphasize Apollo's less positive aspects. The first involves a criticism of the space program from the political left: some critics, both at the time and afterward, deemed Apollo a waste of funds, a project that yielded little at a time when many Americans could have benefited from increased spending on social programs. The second counter-story criticizes Apollo from the right of the political spectrum by focusing on the program as representative of liberal tax-and-spend policies. Finally, a fringe perspective sees in the US Apollo program close ties to all manner of nefarious activities and emphasizes conspiracy theories—of extraterrestrial visitation, abduction, and government complicity—and outright denials of the Apollo Moon landings as products of some deep-seated plot or as part of a larger militarization scheme aimed at world domination. This way of recalling Apollo includes a host of strange and bewildering conspiracies that their adherents believe affected the lives of ordinary Americans in negative ways.

Each of these recollections of Apollo—the mainstream one and those from the left, the right, and the fringes—has its place in the American consciousness. The story that follows explores all four beliefs and how they have evolved in various ways, using each as a touchstone for the ways that Apollo has been remembered over the five decades years since the Moon landings and bringing to the fore the passionate cultural debate over the program that still swirls in the first decades of the 21st century. Each chapter of the book focuses on a major theme in our memories of Apollo, revealing the ways in which it has been seen as a positive endeavor, as well as the ways in which it remains rooted in a time and a place far removed from both our present concerns and our future priorities.

Versions of Reality

Pete Conrad, the sole Apollo astronaut who had received an Ivy League education, spent the rest of his life trying to live it down. Even his entry onto the lunar surface, as part of the Apollo 12 mission on November 19, 1969, was marked by a sort of insistent folksiness. After making a precision landing within 600 feet of a robotic soft lander that had reached the Moon in 1967, he exclaimed to the listening world, in sheer excitement, "Whoopee! Man, that may have been a small one for Neil, but that's a long one for me." Thus he entered the public's perception in a way strikingly different from the way that the astronauts of the first Moon landing had been perceived. While some decried his exclamations as unbecoming of a national representative who was speaking to, and for, the entire world, others enjoyed his words and even saw him as a sort of folk hero. That perspective was reinforced when Conrad, in later recollections, said he was just happy that he hadn't cursed. Brief as this episode was, it suggests the divergent ways in which Apollo has been presented to the public and remembered in the years since humans last set foot on the Moon.

Another story—this one an old baseball joke—can also help us to understand the various ways we have talked about Apollo. Three umpires are discussing how they call balls and strikes behind the plate. One umpire says, "I call them as they are." The second says, "I call them as I see them." The third says, "They ain't nothin' 'til I call them."

The first umpire's view of the nature of reality, what it is, and who decides reflects a premodern, absolutist position: reality exists, he says, beyond the ways that any bystanders might perceive or interpret it. The second umpire's position, one of rationality and modernity, suggests that reality is shaped by perceptions. The third umpire's statement expresses a postmodern, existential belief about reality: what actually happened does not matter all that much, and the thing that *is* truly important is the decision he makes about that reality's meaning. It seems that this last perspective has become the critical element in the ways Americans consider the Moon landings. What we think of Apollo is an intensely personal matter, one predicated on our many idiosyncrasies and perspectives.

Without question, Apollo was for most observers, and perhaps consistently for much of the American public, an epochal event that signaled the opening of a new frontier where a grand visionary future for Americans might be realized. It epitomized, most Americans have consistently believed, what sets the United States apart from the rest of the nations of the world. American exceptionalism reigns in this telling of Apollo, and the program is often depicted as a momentous undertaking in US history, one that must be revered because it shows how successful Americans can be when they try.

Apollo has been consistently viewed as a feel-good triumph for America and its people, an iconic moment of national accomplishment. An astronaut on the Moon saluting the American flag served well as a patriotic symbol, one made all the more powerful because it required Americans to overcome an existential threat from the Soviet Union. Celebrants of Apollo, accordingly, have long argued that returns on investment in the age of space exploration changed Americans' lives for the better. As President Lyndon B. Johnson remarked in response to successes in the Moon race in August 1965, "Somehow the problems which yesterday seemed large and ominous and insoluble today appear much less foreboding." Why should Americans fear problems on Earth, he believed, when they had accomplished so much in space? If we can go to the Moon, why can't we solve our other problems? In this story, the Moon landings demonstrate that we can accomplish anything we set our minds to. After July 20, 1969, "If we can put a man on the Moon, why can't we do X?" became a catchphrase, with the last part of the sentence filled in by any number of ambitious goals. Apollo fits beautifully into a long-standing desire to embrace American awesomeness.

This narrative found expression not just in the United States, but also around the world. One of the objectives of the Apollo program, as noted earlier, was to demonstrate American technological superiority as a means to bring allies into the Western camp during the Cold War, and it achieved this goal. Official congratulations poured in to the US president from other heads of state after the first landing, even as informal ones went to NASA and the astronauts. Every nation that had regular diplomatic relations with the United States sent its best wishes in recognition of the success of the mission. By providing evidence of American technological prowess via the Moon landings, NASA served a valuable foreign policy objective, for it helped keep other nations in America's corner during the Cold War.

A counter-story to this dominant interpretation of American triumph and exceptionality also emerged in the 1960s: some argued that Apollo deserved criticism as a "moondoggle," a wasteful expenditure of federal funds that could have been spent much more effectively elsewhere. For example, Vannevar Bush, a well-respected leading scientist who appreciated the marshaling of the federal government's power in the furtherance of national objectives, wrote to NASA Administrator James E. Webb in April 1963 to voice his concerns about the cost versus the benefits of the Moon program. He asserted that Apollo, "as it has been built up, is not sound," and he expressed concern that it would prove "more expensive than the country can now afford," adding that "its results, while interesting, are secondary to our national welfare."

Sociologist Amitai Etzioni was even more critical. In a reasoned, full-length critique of Apollo in 1964, he deplored the "huge pile of resources" spent on space, "not only in dollars and cents, but the best scientific minds—the best engineering minds were dedicated to the space project." Could not those resources have been better spent on improving the lives of people in modern America? Etzioni bemoaned the nation's penchant for embracing both high technology and unsustainable materialism: "We seek to uphold humanist concerns and a quest for a nobler life under the mounting swell of commercial, mechanical, and mass-media pressure."

Several political leaders in the United States, especially in the Democratic Party, found that support for Apollo clashed with their support of funds for social programs enacted through Great Society legislation. They disparaged Apollo as being both too closely linked to the military-industrial complex and defense spending and too far removed from the ideals of racial, social, and economic justice at the heart of the positive liberal state that most Democrats envisioned. Senators such as J. William Fulbright, Walter Mondale, and William Proxmire challenged the Johnson administration every year over funding for Apollo that they believed could be more effectively used for social programs. Bureau of the Budget Director Charles Schultze worked throughout the mid-1960s to shift funds away from Apollo toward programs such as the war on poverty. Meanwhile Johnson tried to defend Apollo as a part of his Great Society initiatives, arguing that it helped poor Southern communities through its infusion of federal investment into high-technology development programs located there. This proved to be a difficult sell, and the NASA budget declined precipitously throughout the latter half of the 1960s.

This perspective views Apollo as a waste, a missed opportunity to further important and necessary goals in America. Still, the triumphalist narrative of Apollo has been so powerful a memory that today most people in the United States believe that the program enjoyed universally enthusiastic support during the 1960s and that somehow NASA lost its compass thereafter. In fact, at only one point before the Apollo 11 mission, in October 1965, did more than half of the public favor the program. In most public opinion polls of the 1960s, Americans consistently ranked spaceflight near the top of those programs they wished to see cut in the federal budget; a majority ranked the space program as the government initiative most deserving of reduction and said its funding should be redistributed to Social Security, Medicare, and other social programs. While most Americans did not oppose Apollo per se, they certainly questioned spending on it during a time when social problems appeared more pressing.

At the same time, some figures on the American political right during the Apollo era criticized the program as an abuse of federal power. In their view, the federal government should not do much of anything, let alone put humans on the Moon; these critics held a persistently libertarian position that emphasized individual prerogative and freedom over state action. One such critic, former President Dwight D. Eisenhower, believed that empowering NASA to accomplish the Apollo Moon landings was a mistake, remarking in a 1962 article, "Why the great hurry to get to the moon and the planets? We have already demonstrated that in everything except the power of our booster rockets we are leading the world in scientific space exploration. From here on, I think we should proceed in an orderly, scientific way, building one accomplishment on another." He later cautioned that the Moon race "has diverted a disproportionate share of our brain-power and research facilities from equally significant problems, including education and automation." Likewise, in the 1964 presidential election, the Republican candidate, Senator Barry

Goldwater, urged a reduction of the Apollo commitment to pay for national security initiatives.

After the Moon landings came off successfully, however, the American right largely retreated from high-profile criticism of Apollo, and it stayed quiet on the topic until the 1980s, when a full-scale assault on 1960s-era Democratic Great Society initiatives emerged as a central part of the conservative strain that emerged fully during the Reagan years. Conservatives reappraising the '60s—especially its social upheavals and the US defeat in Vietnam—castigated the decade as a time of failure in American politics. Thus a new conservative space policy, as well as a conservative space history, emerged in the 1980s to criticize Apollo.

Nothing expresses this right turn better than the rehabilitation of Eisenhower as president. In reinterpretations by some 1980s historians, including Pulitzer Prize–winning historian Walter A. McDougall (who critiqued Apollo as part of a larger assault on what he believed were "products of the maniacal 1960s"), Eisenhower emerges as the hero of the early space age, seeking to hold down expenditures, refusing to race the Soviet Union into space, and working to maintain traditional balances among policy, economics, and security. As historian of technology Alex Roland has pointed out, "McDougall pictures [Eisenhower] as standing alone against the post-Sputnik stampede, unwilling to hock the crown jewels in a race to the moon, confident that America's security could be guaranteed without a raid on the Treasury, and concerned lest a space race with the Russians jeopardize America's values and freedoms and drag us down to the level of the enemy."

Such linkages of space policy to social policy may seem tenuous at first, but conservative Apollo critics disapproved of the power of the federal government and the state system to "intrude" into individual lives. McDougall—to further examine one exemplar of such critiques—also argued that the space age had given birth to a state of "perpetual technological revolution" due to the industry that arose to support Apollo's incredibly complex set of machines. He focused upon the role of the government as a promoter of technological progress and argued that the United States, in its drive to respond to the Soviet challenge, had recreated the same type of command technocracy that the Soviets themselves had instituted. He saw this as a detriment to American society as a whole, concluding that the Moon race had led to a cooptation of individual liberty in the pursuit of state power and saying that any assertion of state authority represented an equal degradation of individual freedom:

> In these years the fundamental relationship between the government and new technology changed as never before in history. No longer did state and society react to new tools and methods, adjusting, regulating, or encouraging their spontaneous development. Rather, states took upon themselves the primary responsibility for generating new technology. This has meant that to the extent revolutionary technologies have profound second-order consequences in the domestic life of societies, by forcing new technologies, all governments have become revolutionary, whatever their reasons or ideological pretensions.

Once institutionalized, technocracy has not gone away, a fact that led McDougall to conclude that Apollo was enormously costly to the nation, and not just in public treasure. Like other critics of Apollo on the right, he bemoaned government activism and government power, something that conservatives tended to challenge even though most Americans of the era accepted at face value the benign nature of that power.

Finally, there is the third strand of the Apollo story. Almost from the point of the first Apollo missions, a small group of Americans has denied that the race to the Moon—and particularly the Moon landings—took place at all. This group seems to be expanding, and its narrative is rising in importance as the events of Apollo recede into history, aided by a younger generation with no personal memory of what happened in the space-race era and for whom distrust of government runs high. Only 18 percent of all Americans in December 2017 said they trusted the government in Washington to do what is right "always" or "most of the time," and among those younger than 40, such trust was lower still. Jaded by many other government scandals, and by real conspiracies such as those that took the United States into war in Iraq and fostered an economic crisis in 2008, some members of society have found themselves susceptible to the myriad Moon hoax advocates.

Denying the Moon landings invokes a response of "crank" and "crackpot" from most listeners; indeed, these conspiracy theories deserve disdain, and most Americans, surveys consistently show, reject them. But their adherents can be viewed as part of a postmodern interpretation of history: at some level, Americans today, like the umpire who says, "They ain't nothin' 'til I call them," do not see absolutes in our history. Instead, we see everything as constructed. Versions of the past have replaced earlier versions that once *seemed* to be true, even as questions are raised about who has the right—not to mention the power—to interpret the past. Many timeworn narratives about America's history have been dismantled or revised in recent decades, from the absence (or even celebration) of the genocide of Native Americans in stories of the country's founding to historian Frederick Jackson Turner's 1893 frontier thesis, which argued that the national character was forged in westward expansion and which held sway for many decades before it was critiqued and abandoned in the late 20th century. Meanwhile other conspiracy theories have become major elements of the memory of the nation—for example, the belief that John F. Kennedy was assassinated by some nefarious network involving key members of the national security establishment rather than by a single gunman. It is conceivable that such an erroneous reevaluation might one day reshape our view of the Moon landings, too.

In the context of Apollo's history, the duels among its four primary "stories"—the dominant celebratory account of the program; critiques from the left and from the right; and the conspiracy theories—have been, in essence, a battle for control of national memory. Was Apollo an exemplar of American uniqueness, a bright and shining moment in the nation's history to be remembered that way for all time? Or was it a waste of time, energy, and resources? Was it a mistake? Or was it all a hoax? To return to the story of the umpires: do we call it as it is, call it as we see it, or claim that it's nothing until we've called it in the first place?

$$\boxed{2}$$

A Moment in Time

Ted Sorensen awoke early on the morning of Thursday, May 25, 1961. As President John F. Kennedy's chief speechwriter, senior counselor, and "intellectual blood bank," the 33-year-old Sorensen was worried about the words the president would speak that day at a joint address to Congress entitled "Urgent National Needs." The speech needed more work, Sorensen thought; it did not soar the way other JFK speeches had done. For one thing, it failed to respond directly to what had transpired just the preceding month, when Yuri Gagarin of the USSR became the first human being to travel into outer space, completing an orbit of the globe aboard *Vostok 1*. More important, Sorensen thought, the speech didn't contain memorable lines that would be recited years later. There was no "Ask not what your country can do for you; ask what you can do for your country," a line that had inspired so many people at JFK's inauguration earlier that year.

All day Sorensen made changes to the speech. It held the important themes and the principles that the president wanted to present concerning military preparedness, economic improvements, anti-recessionary measures, civil defense, the possibility of lessening tensions with the Soviet Union, and assistance both to allies and to the less fortunate of the world, as well as the value of telling the blessings of liberty and capitalism around the globe. How might Sorensen make the speech unforgettable? He consulted some of the greatest presidential speeches in American history—the Gettysburg Address, George Washington's first inaugural address, Lincoln's second inaugural, and Franklin D. Roosevelt's first inaugural, with its ringing phrase "The only thing we have to fear is fear itself"—in search of an answer. Sorensen sought lofty, expansive, witty, and muscular imagery and a spiraling sense of challenge and commitment. He did not find a satisfying answer.

Sorensen never betrayed his concerns to the president, and if JFK had doubts about the speech before he delivered it, he never voiced them either. Both knew that the most significant part of the lengthy speech, which was focused around both terrifying Cold War concerns (such as nuclear holocaust) and more mundane topics (such as the need for more civil-defense training funds), was an announcement that NASA would undertake a human lunar mission by the end of the decade. Sorensen had ensured that that section contained the highest rhetoric. Before Congress that afternoon, Kennedy spoke his words: "I believe that this nation should commit itself to achieving the goal, before this decade is out, of landing a man on the moon and returning him safely to the earth." He couched this heady objective as a Cold War initiative:

> If we are to win the battle that is now going on around the world between freedom and tyranny, the dramatic achievements in space [by the USSR] which occurred in recent weeks should have made clear to us all, as did the Sputnik in 1957, the impact of this adventure on the minds of men everywhere who are attempting to make a determination of which road they should take. . . . We go into space because whatever mankind must undertake, free men must fully share.

He concluded, "No single space project in this period will be more impressive to mankind, or more important for the long-range exploration of space; and none will be so difficult or expensive to accomplish."

JFK recognized that this effort would require a significant investment from the public treasury, and, even as he delivered his promise of "landing a man on the moon and returning him safely to the earth," he recognized that this funding would be a point of controversy among both fiscally conservative Americans and socially liberal ones. He knew that those on the left would bemoan the expenditure because those same funds could be used to advantage in a succession of social programs, and he knew that those on the right, like Dwight D. Eisenhower, thought that any funding for space should be redirected toward national security priorities. Kennedy sought not only to assuage both sets of critics but also to caution that they must agree that the lunar effort would be appropriate.

Kennedy's concern about successfully selling this initiative to Congress was immediately apparent. Sorensen recalled that as Kennedy delivered the speech, "the President looked strained in his effort to win them over." As Sorensen rode back to the White House in the president's limousine, Kennedy told him that he was disappointed in the congressional response to his speech, which he found "skeptical, if not hostile," and he thought that "his request was being received with stunned doubt and disbelief." He feared that Congress would not approve the budget increases necessary for NASA to reach the Moon.

Sorensen responded that all would be fine regardless of "the routine applause" with which Congress had greeted the Moon-landing initiative; more practically, regardless of any lukewarm response, Lyndon B. Johnson, the current vice president and a master of the Congress, would line up the members needed to support the effort. Sorensen was correct in his predictions. LBJ had spent the six weeks before Kennedy's speech assessing possibilities, corralling members of Congress, and reining in NASA engineers. He asked everyone what was technically and politically feasible; he horse-traded and cajoled; sometimes he threatened and bullied to gain support for NASA. LBJ was a difficult person to resist; as Senator Richard Russell (D-GA) once famously said of him, "That man will twist your arm off at the shoulder and beat your head in with it."

In his speech, Kennedy had set in motion an effort that eventually ended with Americans' reaching the Moon in 1969. Yet his decision to undertake the program is best understood as occurring at a specific moment in time: in other circumstances, and at other times in his presidency, JFK might have made a different choice, but his Moon decision has taken on a mythical significance as we reflect on it from the distance of many decades later. Like the Kennedy administration itself, JFK's decision in favor of Apollo has been almost universally celebrated as an example of both excellent presidential leadership and outstanding public policy formulation. In this, it echoes other assessments of Kennedy's "Camelot," the nickname that still casts its rosy spell over views of his administration. Seeing his decision in this light fits well into the dominant story of the space program, which tells of a nation positively embracing challenges, persevering, and ultimately triumphing in a race with the Soviet Union.

When Senator John F. Kennedy was running for president as the Democratic candidate in 1960, with congressional wizard Johnson as his running mate, he effectively used the slogan "Let's get this country moving again." Erroneously, but with all the expedience of high-stakes politics, Kennedy charged Eisenhower's Republican administration (and by proxy his opponent, Vice President Richard M. Nixon) with doing nothing about the myriad Cold War issues that had festered throughout the 1950s. He especially criticized Eisenhower's response to the supposed "missile gap," in which the United States presumably lagged far behind the Soviet Union in intercontinental ballistic missile (ICBM) technology. Few understood then (or now) that there had never been an actual missile gap during the Eisenhower administration. In fact, Eisenhower put considerable effort into developing an appropriate policy vis-à-vis the Soviet Union's possible ballistic missile threat and made a sizable investment in reconnaissance satellites and rocket technology even before Sputnik. The billions expended on the US guided missile program would, he believed, more than pay off, and indeed the investment eventually yielded not only the ballistic missile technology that served the country throughout the Cold War but also much of the orbital launch capability that NASA would require. In the wake of Sputnik, as the American space agency was being established, Eisenhower set aside his initial instinct to place NASA under military control and agreed that it should be a civilian agency. He also approved *Explorer 1*, the first US mission in space, a satellite launched only four months after Sputnik as part of the 1957–58 International Geophysical Year, and instructed NASA to undertake Project Mercury with the intention to find out whether human spaceflight was feasible (although he disapproved of NASA's plans for a manned Moon mission). Thus it is incorrect to characterize him as a president who gave short shrift to space issues, at least as they related to national security concerns.

These facts did nothing to dissuade Kennedy from using the supposed missile gap to achieve his political ends. He found it a winning strategy against Nixon, who tried to defend the Eisenhower approach to space without giving away what the United States actually knew about Soviet space capabilities from intelligence sources. Hitting the missile gap hard in the early part of the presidential campaign, Kennedy returned to it just before the election, explicitly chastising Eisenhower in major speeches in October and November for allowing the Soviets to rush ahead in rocket technology. This strategy might have meant the difference in an election that seemed to revolve around Cold War issues. When the results came in, Kennedy gained the presidency by a narrow margin of 118,550 out of more than 68 million popular votes cast.

Kennedy's use of space as an issue in the 1960 campaign did not necessarily mean that he cared intrinsically about the subject, however, and his actions during his first 100 days in office demonstrated that basic disinterest. "Of all the major problems facing Kennedy when he came into office," wrote presidential commentator and journalist Hugh Sidey, "he probably knew and understood least about space." Kennedy listened to his new science advisor, Jerome B. Wiesner of the Massachusetts Institute of Technology, who urged that the new administration deemphasize human spaceflight initiatives for practical reasons: spaceflight was dangerous and unlikely to yield valuable scientific results, he argued. Wiesner specifically cautioned Kennedy about the hyperbole associated with the very public human spaceflight agenda that the Eisenhower administration inaugurated in the aftermath of Sputnik and the creation of NASA. "Indeed, by having placed the highest national priority on the Mercury program, we have strengthened the popular belief that [placing a] man in space is the most important aim of our non-military space effort," Wiesner wrote to the president-elect shortly before Kennedy's inauguration. "The manner in which this program has been publicized in our press has further crystallized such belief."

Accordingly, Kennedy's earliest pronouncements regarding civilian space activity distanced the new administration from any bold spaceflight endeavors. Had the balance of power and prestige between the United States and the Soviet Union remained stable in the spring of 1961, it is quite possible that Kennedy never would have advanced his Moon program, and American space efforts might have taken a radically different course. Kennedy changed his mind about engaging in a space race mostly in response to two important events that embarrassed him: the flight of Soviet cosmonaut Yuri Gagarin on April 12, 1961, and the abortive Bay of Pigs invasion of Cuba, designed to overthrow the revolutionary government of Fidel Castro, just a few days later.

Although the Bay of Pigs invasion was never mentioned explicitly as a reason for stepping up US efforts in space, the international situation certainly played a role as Kennedy scrambled to recover a measure of both national and personal dignity. T. Keith Glennan, who had been NASA administrator under Eisenhower, immediately perceived the invasion and the Gagarin flight as the seminal events leading to Kennedy's announcement of the Apollo decision, confiding in his diary, "In the aftermath of that [Bay of Pigs] fiasco, and because of the successful orbiting of astronauts by the Soviet Union, it is my opinion that Mr. Kennedy asked for a reevaluation of the nation's space program."

Two days after the Gagarin flight, in fact, Kennedy discussed the possibility of a lunar landing program with NASA Administrator James E. Webb, but he felt that even the NASA head's conservative estimate that it would cost more than $20 billion was too steep, and the president delayed his decision. A week later, at the time of the Bay of Pigs invasion, Kennedy called Johnson to the White House to discuss strategy for catching up with the Soviets in space. It is likely that one of the specific programs that Kennedy asked Johnson to consider was a lunar landing program, for the next day he followed up with a memorandum to the vice president: "Do we have a chance of beating the Soviets by putting a laboratory in space, or by a trip around the moon, or by a rocket to go to the moon and back with a man? Is there any other space program that promises dramatic results in which we could win?"

This memo makes it clear that Kennedy, even before he received the results of Johnson's inquiry, already had formulated a pretty good idea of what he wanted to do in space. He said at a press conference on April 21 that he was leaning toward committing the nation to a large-scale project to land Americans on the Moon. "If we can get to the moon before the Russians, then we should," he stated, adding that he had asked his vice president to review options for the space program. (This was the only time that Kennedy said anything in public about a lunar landing program until he officially unveiled the plan in his congressional address of May 25.) This statement demonstrates that he essentially saw the Moon initiative as a response to competition between the United States and the Soviet Union. For Kennedy, the lunar landing program was a strategic decision directed at advancing the far-flung interests of the United States in the international arena and recapturing the prestige that the nation had lost in the wake of Soviet triumphs and American failures in the tense Cold War environment of the spring of 1961.

Kennedy's announcement of the program proved much more easily accepted than he had anticipated. Johnson, as Sorensen had foreseen, knew how to generate and sustain the political support necessary to obtain the resources that would make Apollo a reality. With this decision, Eisenhower's approach to space—with its emphasis upon military, satellite, and launcher technology and utilitarian science—disappeared. A spectacular space adventure that drew fire from both the left and the right, but strong support from the center, replaced it. The fact that his space proposals later sped through Congress may have surprised even Kennedy, however. As he signed the bill authorizing his space initiatives on July 21, he pointed out, with pride, the "overwhelming support by members of both parties."

Congressional debate on the NASA budget in 1961 proved perfunctory—the vote did not even require a roll call—and the agency found itself pressed to expend the funds committed to it during the early days of Apollo. Like most political decisions, at least in the US experience, the decision to carry out Project Apollo was an effort to deal with an unsatisfactory situation: world perception of Soviet leadership in space and technology. As such, Apollo was a remedial action ministering to a variety of political and emotional needs, and it addressed those needs very well. In announcing Apollo, Kennedy put the world on notice that the United States would not take a back seat to its superpower rival. It was an effective symbol, just as Kennedy had intended.

It did not remain one, however. Within a year, attacks on the program from both the left and the right were asserting that Apollo's costs were greater than their worth and should be curtailed in favor of other spending priorities. From 1962 onward, Congress challenged the cost of the program every year, and NASA officials had to scramble to preserve the effort. As it turned out, James Webb, the NASA administrator, was able to sustain Apollo's momentum throughout the decade largely because of his personal rapport with key members of Congress and with Johnson, who became president in November 1963. Accordingly, the space agency's annual budget increased from $500 million in 1960, accounting for less than .05 percent of the federal budget that year, to a high of $5.2 billion in 1965. (A comparable percentage of the $2.1 trillion federal budget in fiscal year 2017 would have equaled more than $85 billion for NASA, whereas the agency's actual budget that year was just $19.4 billion.) NASA's budget began to decline in 1966 and continued its downward trend until 1975. The situation got so bad in 1970 that NASA's next-year budget forced the agency's leadership to cancel Apollo missions 18, 19, and 20. Apart from a few years in the middle of the Apollo era, the NASA budget has hovered at less than 1 percent of all money expended by the US Treasury annually.

As successful as Kennedy's 1961 decision appears in retrospect, he made it recognizing its inherent risks. Possibilities for failure abounded, as did those for derogation of American prestige in the international arena and political ridicule and perhaps defeat in the 1964 election for himself. Kennedy worried aloud about these pitfalls in an Oval Office meeting with Webb on September 18, 1963. "It's become a political struggle now," Kennedy said. "I don't think the space program has much political positives." He vowed to continue, but he was already thinking about how he would justify the Moon-landing program when he sought reelection in 1964. One tactic he believed would work was to link it to national security and head-to-head competition with the Soviet Union. Had he lived, no doubt Kennedy would have made that case on the reelection trail. With his assassination in November 1963, however, it fell to Lyndon Johnson to carry the program forward. Johnson remained committed to Apollo despite pressures to spend those funds elsewhere.

Since the Apollo program, many investigators have sought to understand why Kennedy decided to go to the Moon. Most espouse the interpretation suggested above: that Kennedy made a single, rational, pragmatic choice to undertake the US sprint to the Moon as a means to increase national prestige as the country competed with the Soviet Union during the height of the Cold War. The president and his advisors, in this view, undertook a deliberate, reasonable, judicious, and logical process to define the problem, analyze the situation, develop a response, and achieve a consensus for action. Their timeline progressed from point to point, with no cul-de-sacs and few detours along their path from problem definition to sensible decision. This "rational choice" argument begins and ends with the assertion that JFK's space policy was a relic of the Cold War struggle between the United States and the Soviet Union, and it revolves around the question of international power.

In this view, Apollo represents a clear understanding of the competition between the world's two superpowers to win people's minds over to a specific economic and political system. The Apollo program was nothing less than the "moral equivalent of war," according to supporters such as Johnson. It sought to weaken the Soviet Union while enhancing the United States. It was neat and tidy.

There is much to recommend this understanding of the Apollo decision, and it is the one that dominates most American recollections of the space race. A year after the decision, Kennedy defended it on the grounds that there were only three ways that the United States could effectively compete with the Soviet Union: militarily, economically, and technologically. In 1962, he made the case to administration officials that he'd eliminated the first option because no one desired nuclear war. The second option was unattractive because it would take a long time for a clear economic winner to emerge. Apollo presented a logical third option because the technological systems necessary to land on the Moon did not yet exist anywhere, and thus the Americans, already racing to meet the Soviet edge in launch vehicle technology, could compete on an equal basis with the Soviets. With commitment and diligence, JFK believed, the United States could win a fair race to the Moon.

In this sense, Kennedy's Apollo decision was a politically pragmatic one that simultaneously solved several significant problems for his administration. At the same time, the flaw in this interpretation is its unwavering belief that individuals—especially groups of individuals, even competing ones—actually do logically assess situations and respond with reasonable consensus actions. Since virtually nothing in human existence is done solely on a rational basis, this is a very difficult belief to accept.

A second, more speculative explanation for the Apollo decision suggests that Kennedy's privileged upbringing by a stern father fostered aggressive tendencies that affected his decision making, causing him to take a more combative approach toward the Soviet Union than was required and necessitating his "winning" at whatever challenge came his way. At some level, Kennedy may have even created crisis situations in which he could reaffirm his quintessential masculinity and enhance his own dominance over everyone and everything. Most of these analyses depict JFK in an unfavorable light and focus on the tendencies toward overarching competitiveness, general recklessness, and Machiavellian ambition instilled in all the sons of Joseph P. Kennedy. These character studies view Kennedy as an individual who needed to dominate people and who unconsciously, or in some cases deliberately, created situations calculated to demonstrate his mastery. His treatment of women (he was an ardent and destructive philanderer) might support such an argument, as might his brutal competition with men in sports, business, and politics. Presidential biographer Richard Reeves's assessment—"He was a hard man, casually cruel"—is telling, and he has written that he was unsure whether he even liked his subject, a man so aggressive that he "turned over checker and backgammon boards when he realized he might lose a game."

It may well have been that Kennedy's flaws, as much as anything else, pointed his direction toward the Apollo decision in 1961. In contrast to Eisenhower, who had refused to fall prey to public hysteria after the Sputnik launches in 1957—saying at one point that he did not see the need to pursue a Moon program, since the United States had no enemies on the Moon—Kennedy fanned the flames of crisis that sprang up in the aftermath of Gagarin's flight. There is not much doubt that Kennedy came into office having thought very little about how the United States would deal with the Soviet Union. He had the patina of the Cold Warrior common to the era, but beyond that generic anticommunism, he did not display much of what might be termed strategic thinking about the superpower rivalry. Harold Macmillan, prime minister of the United Kingdom, observed this fact in his diary after meeting Kennedy in March 1961, saying that while he was unimpressed with the new American president's strategic thinking, he found JFK masterful at making decisions under pressure, "a sort of Duke of Wellington of America." Kennedy had the ability to ride out almost any storm through a series of expedient decisions that often contradicted one another but proved helpful in responding to current issues. He did so not only in his responses to the Gagarin flight and the Bay of Pigs debacle in April 1961, but also in the wake of the Berlin crisis of 1961, the monk Thich Quang Duc's self-immolation in Vietnam in June 1962, the Cuban missile crisis in October 1962, and the Birmingham civil rights march, infamous for attacks on protestors by police and their dogs, in May 1963. There is little in any of his responses to these crises to suggest a strategic vision or value system beyond personal muddling through.

The missile crisis is a case in point. Kennedy's handling of the placement of Soviet ballistic missiles in Cuba, a 13-day (October 16–28) confrontation that is often considered the closest the Cold War came to escalating into a full-scale nuclear war, is usually viewed as masterful. Robert M. Kennedy's account *Thirteen Days* (1969) celebrates the administration's combination of saber-rattling and cautious diplomacy and emphasizes rational actions on the part of JFK and his crisis team. It says little, though, about Kennedy's initial leak of the idea of negotiating a public trade of Soviet ballistic missiles in Cuba for American Jupiter ballistic missiles aimed at the USSR from Turkey. When that was perceived as American weakness, JFK disowned the strategy. He also initially wanted to attack the launcher sites in Cuba, but cooler heads urged restraint lest the attack lead to nuclear war. An approach that included a blockade of Russian materiel entering Cuba and back-channel communication eventually led to a closet deal to remove the Soviet missiles, followed within a few months by the Jupiter missiles' being dismantled secretly as well. Kennedy wavered throughout the crisis from one approach to another, with no firm decision-making in evidence. The strength of the advisors around him, as well as Soviet moderates, led to a successful lessoning of tensions.

Kennedy had no refined strategy for how to win the Cold War, and thus he had an accordingly greater capacity to view each problem as if he were in a death match. For him, each confrontation with the Soviet Union took on spectacular proportions and desperate characteristics. Indeed, had Eisenhower been in office in 1961, it is doubtful that he would have responded to international setbacks with a similar lunar landing decision. Instead, he probably would have sought to reassure those stampeded by Soviet achievements and explain carefully the long-term approach NASA was taking toward space exploration.

Instead of taking such a long view, Kennedy exploited fears about supposed Soviet strength in space, which he juxtaposed against supposed American weakness, and he responded with a lunar landing decision both spectacular in its achievement and outrageous in its cost. Kennedy, because of his competitive nature, was apparently anxious to metaphorically strap on his six-guns and shoot it out with Soviet Premier Nikita Khrushchev, in keeping with the Western television and films so popular during the early 1960s. He recognized that this might not be the most effective way to deal with the Soviet Union, but he kept at it.

JFK's vacillation was also evident in space activities. While he began the space race as a Cold Warrior, Kennedy also pursued a cooperative path with the Soviet Union in space, certainly something never hinted at in the subsequent Johnson era. In his inaugural address in January 1961, Kennedy spoke directly to Khrushchev, asking him to cooperate in exploring "the stars." In his first State of the Union address, delivered 10 days later, he asked the Soviet Union "to join us in developing a weather prediction program, in a new communications satellite program, and in preparation for probing the distant planets of Mars and Venus, probes which may someday unlock the deepest secrets of the Universe." Even after Gagarin and the Bay of Pigs, in the weeks preceding his May announcement of Apollo, Kennedy had his brother Robert quietly assess the Soviet leadership's inclinations toward taking a cooperative approach to space exploration. In addition, NASA Deputy Administrator Hugh L. Dryden undertook a series of talks with Soviet academician Anatoli A. Blagonravov about joint ventures in space. They reached agreement on several collaborative efforts, such as exchanges of scientific data, but not on major human missions. Kennedy also instructed Wiesner, his science advisor, to convene a panel with representatives of NASA and the President's Science Advisory Committee that would generate ideas for cooperation with the Soviets, including setting up an international lunar base. In a memo written the same day as JFK's Apollo speech, Eugene Skolnikoff, who was on Wiesner's staff, proposed that "we should offer the Soviets a range of choice as to the degree and scope of cooperation." As Ted Sorensen later remarked, "It is no secret that Kennedy would have preferred to cooperate with the Soviets on space exploration."

Within two weeks of giving his bold May 25 speech, Kennedy met Khrushchev at a summit in Vienna and proposed that Apollo be a joint mission with the Soviets. The Soviet leader reportedly first said no, then replied, "Why not?" Then he changed his mind again, saying that disarmament was a prerequisite for American-Soviet cooperation in space. Later, on September 20, 1963, Kennedy spoke before the United Nations and again proposed a joint human mission to the Moon.

In public, the Soviet Union was noncommittal. *Pravda*, for example, dismissed the 1963 proposal as premature. Some have suggested that Khrushchev viewed the American offer as a ploy to open up Soviet society

and compromise Soviet technology. Although these efforts did not produce any space agreements, Kennedy's pursuit of various forms of space
cooperation until his death suggests that he was unsure whether going it
alone in the Apollo program was the best course.

A tape of a White House meeting between Kennedy and the NASA
administrator on November 21, 1962, a month after the Cuban missile
crisis, does suggest that Kennedy believed cooperation with the Soviet
Union was a viable course. The cost of acting unilaterally, he said, had
proven greater than anyone envisioned in 1961. He questioned the viability of such an expansive space agenda except as a Cold War objective.
When asked to support a broad range of spaceflight activities on the
basis of their scientific merits, Kennedy refused, telling Webb, "I am not
that interested in space." He was spending so much money on Apollo,
he insisted, mostly because of its importance in the Cold War rivalry with
the Soviet Union.

Since JFK's Apollo decision, many space enthusiasts have scrutinized
his lunar decision almost to the point of banality. They routinely draw the
lesson that direction of a major goal in space from the president should
allow NASA to proceed without any opposition. The belief that Congress
and other parts of the executive branch, as well as the public, would get
behind a perceived visionary presidential announcement on space has
been repeatedly proven naïve in later instances when a president made
bold statements about space exploration. Something most NASA officials
did not understand at the time of the Moon landing in 1969, or for many
years thereafter, was that the situation that had given rise to the Apollo
decision had not been an ordinary one, and that those circumstances
would not be repeated. Kennedy's decree was, in reality, an anomaly in
the national decision-making process. It was carried out only because it
relied on the idea of presidential prerogatives and because it was sold to
Congress on the basis of its national defense implications.

Not until the 1989 debate over the stillborn Space Exploration
Initiative (SEI), promoted by President George H. W. Bush, did the space
community experience their inevitable confrontation with reality. At this
point, they at last came to appreciate that presidential support for a space
endeavor does not automatically guarantee its success. The demise of SEI
forced space exploration champions to question long-held assumptions
about presidential omnipotence, and faith in the ability of presidential

commitments to free space policy from the constraints of Washington politics declined thereafter. Expensive space efforts such as SEI have proved to be extremely tough sells.

Still, the symbolism of Kennedy's Apollo commitment has held special appeal for true believers in space exploration from the time of his May 25, 1961, speech to the present day. To them, the lunar decision suggests that space exploration deserves special treatment within the American political system. The decision to go to the Moon implied that a president could overcome partisan divisions and lead the nation to great accomplishments, if only the objective were properly framed. Many argue that the subsequent ills of the space program can be traced to the unwillingness of later presidents to make Apollo-like public commitments.

If Ted Sorensen felt that he did not succeed in capturing the imagination of the public with his text for JFK's announcement of the Moon-landing endeavor, he made up for it when the president delivered a stunning speech on Apollo at Rice University's stadium on September 12, 1962, by which time the program was well underway and NASA was establishing its Manned Spacecraft Center in Houston:

> We set sail on this new sea because there is new knowledge to be gained, and new rights to be won, and they must be won and used for the progress of all people. . . . *We choose to go to the Moon!* We choose to go to the Moon in this decade and do the other things, *not* because they are easy, *but because they are hard;* because *that goal* will serve to organize and measure the best of our energies and skills, because *that challenge* is one that we are willing to accept, one we are unwilling to postpone, and *one we intend to win.*

Sorensen offered Kennedy the rhetoric he needed to characterize space as a beckoning frontier. Apollo was positioned as a grand endeavor in a historically urgent period, and the lunar program as a celebration of America's pioneering heritage extended to the Moon. The Rice speech set the Moon-landing effort in a new context, and JFK reveled in it. It was one of his greatest speeches, and he knew it.

His words that day cast in concrete a perception, already becoming lodged in memory, of a visionary president calling America to a higher place. He made Apollo sound like a great popular triumph, a far-sighted national commitment willingly accepted by the overall population. This is an important conception, for without the active agreement of political leaders and the public's acceptance, no expansive space effort can be sustained for any length of time. Broad public enthusiasm was thereafter perceived as something that had to be gained for any large new space exploration effort. Among space exploration advocates even today, one repeatedly hears a chorus lamenting the supposedly lukewarm popular support that their work has received since the Kennedy decision: "If only our current efforts had the same level of commitment enjoyed by Apollo, all would be well," they say, following with a heavy sigh.

While there is reason to believe this statement at some basic level, it is at best a simplistic and ultimately unsatisfactory observation. Since the mid-1960s, as noted earlier, there has been neither overwhelming US public support for lavish space exploration funding nor overwhelming disapproval of such efforts. Instead, support for space funding has remained remarkably stable since that time, with approximately 80 percent in favor of *modest* expenditures for space flight. In the summer of 1965, a third of the nation favored cutting the space budget, while only 16 percent wanted to increase it. Over the next three and a half years, the number in favor of cutting space spending went up to 40 percent, with those preferring an increase dropping to 14 percent. At the end of 1967, the *New York Times* reported that a poll conducted in six American cities showed five other public issues holding priority over efforts in outer space, including air and water pollution, job training for unskilled workers, national beautification, and poverty.

Looking back years later, it might seem surprising that NASA's effort to land on the Moon during the 1960s never garnered great public support and that JFK vacillated about it as he considered his reelection bid in 1963; it now seems such a triumph that it is hard to accept that enthusiasm for the program at the time was always tempered by its high price tag. Because of this, it is tempting to ask the question "What might have happened to the Apollo program had JFK not fallen to an assassin's bullet on November 22, 1963?" Would it have continued as it did, or might it have followed some other course? We shall never know. JFK did fall in 1963, and Apollo did reach the Moon in 1969. The flight of Apollo 11 has become historical fact, and one of the greatest achievements of humankind.

At a basic level, the president's Apollo decision was to the United States what the pharaohs' determination to build the pyramids was to Egypt. But the circumstances of the decision were unique, wrapped up in the singular crisis atmosphere of the spring of 1961, the personality of John F. Kennedy, and the rivalry of the two superpowers in space. The decision was enormously important as a public effort, and for the rich scientific harvest of Apollo space flights that it yielded, but it was the product of a specific time and set of circumstances that US space exploration advocates should not expect to see again.

How might those recalling Apollo most accurately view that moment in time, now that nearly six decades have passed? Is it best remembered as an episode in which a great American leader stepped forward with a bold initiative that would forever characterize America as an exceptional nation committed to high ideals? Was it an exemplar of crisis management? Or was it something else? The answer might be: all of the above.

$$\boxed{3}$$

The Most Powerful
Technology Ever Conceived

A standard story, perhaps apocryphal, told at the time of Apollo concerned the reliability of the technology that would take astronauts to the Moon. NASA wanted to build the spacecraft and rockets used in the human spaceflight program so well that they would work effectively 99.99 percent of the time. In their rigorous engineering culture, NASA engineers reasoned, there was a chance of catastrophic failure—in which an astronaut fatality would occur—only once in every 1,000 flights. Safety to "four nines" became the agency's watchword. Everyone associated with Apollo strenuously tried to achieve such a standard, and while the agency might not have met its own benchmark, NASA's safety culture has been the envy of other technical organizations in the world throughout its history.

In the early years of Apollo, the story goes, a group of politicians visited Wernher von Braun at the George C. Marshall Space Flight Center in Huntsville, Alabama, where his team was building the mighty Saturn V Moon rocket. Following discussion of NASA's four nines goal, they asked him about the reliability of the rocket. In response, Von Braun introduced four of his senior lieutenants, all of whom had come from Germany with him at the end of World War II, and put to them a simple question: "Is the rocket going to fail?"

In turn, each of the four engineers responded, "Nein," the German word for *no*, which is pronounced exactly like the American word *nine*.

Laughing, Von Braun said, "There you have it: the Saturn V is safe to four nines."

Whether true or not, this story points up an important reality about the Apollo program. Although it used newly developed technology, it also built on knowledge gained through years of research, development, and testing. While the objective of 99.99 percent reliability exhibited a striking amount of hubris (and never actually has been achieved in any spaceflight program), it was a noble aim for near-perfection.

Recollections of the Apollo program's technology still lead many to express wonder at the sophisticated technical competence that made the Moon landings possible, and the genius of those who built the rockets and spacecraft that carried Americans into space. Around the Moon landing's 40th anniversary, Farouk el-Baz, a scientist who had worked on the program, expressed well this sense of awe, saying, "Oh, the Apollo program! It was a unique effort altogether. . . . I still look at that time with wonder," and bemoaning the fact that "the Apollo spirit of innovation and can-do attitude did not last long."

The technology used to go to the Moon in the 1960s seems unsophisticated by 21st-century standards. Many people are surprised to learn that there is more computing power in a modern smartphone than in all the Apollo computing systems together, both the onboard ones and those on Earth. Others are surprised to learn about the fine-grained level of the technology that spaceflight demanded of NASA's engineers. Even something as simple as writing in space required the development of a new type of pen, made by the Fisher Co., with pressurized ink that could flow even in a weightless environment.

American belief in the technical virtuosity of NASA, an agency that seemingly could accomplish any task assigned it, can be traced directly to the experience of Apollo and its legacy of achievement. Apollo's accomplishments still have resonance, as pointed out earlier, in the often-used question "If we can put a man on the Moon, why can't we do *X*?" (for *X*, substitute the task of your choice). NASA's success in reaching the Moon established a powerful conception in American culture that one could make virtually any demand and the space agency would deliver. Despite tragedies along the way, including the near-disaster of Apollo 13 and the very public *Challenger* and *Columbia* accidents, which together killed 14 astronauts, the clear majority of the public is still convinced that NASA has the capability to succeed at whatever it attempts. The Moon landings established that image in the American mind, and it has been difficult to tarnish despite the space agency's failures after Apollo.

Memories of Apollo's success have also been colored by concerns over a nebulous sense of decline present in numerous parts of modern society. Many Americans have expressed a desire to recapture the spirited and genuine technological virtuosity that existed in the 1960s. El-Baz said, in recalling the program, "I believe that my generation has failed the American people in one respect. We considered Apollo as an enormous challenge and a singular goal. To us, it was the end game. We knew that nothing like it ever happened in the past and behaved as if it would have no equal in the future."

The technology required to reach the Moon was certainly more complex than anything ever attempted before, but its outlines were firmly understood at the time that the program began. NASA engineers reasoned, first, that they needed a more powerful rocket with a larger payload capacity than any envisioned before. As a second priority, they recognized the need to develop a spacecraft that could preserve the life of fragile human beings for at least two weeks. The spacecraft had two elements: a vehicle, akin to a submarine, that could protect astronauts while they traveled in space, and a spacesuit that would allow an astronaut to perform tasks outside the larger vehicle. Third, the engineers needed to develop some type of landing craft that could operate in the Moon environment, far different from that of anywhere on Earth. Finally, they needed to develop the technologies necessary for guidance, control, communication, and navigation. In every case, planners at NASA understood the nature of the technical challenges before them, which enabled them to chart a reasonable and well-defined technology development course for overcoming them.

Notwithstanding, many at NASA questioned whether the space agency could be successful in meeting Kennedy's mandate to land an American on the Moon by the end of the decade. Robert Gilruth, the director of the NASA Space Task Group charged with undertaking human spaceflight, recalled that he was initially "aghast" at the lunar landing goal, given what accomplishing it would require. Yet in the end, Gilruth remarked in 1986, the Apollo program proved to be "pretty straightforward" despite all the complex technologies that had to be created, with specifications both exacting and redundant because they were "single-point failures in the sequence."

For the first challenge, that of rocketry, NASA inherited an effort to develop the Saturn family of boosters in 1960 when it acquired the Von Braun team of rocketeers at the Army Ballistic Missile Agency in

Huntsville. By this time, Von Braun's engineers already were hard at work on the first-generation Saturn I launch vehicle, which was then envisioned as weaponry: namely, a new type of ballistic missile. Fueled by a combination of liquid oxygen (LOX) and RP-1 (a version of kerosene), the Saturn I could generate an impressive thrust of 205,000 pounds, more than any other rocket of the early 1960s. The group was also working on a second stage, known as the Centaur, that used a fuel mixture of LOX and liquid hydrogen (LH2), a revolutionary combination that generated a greater ratio of thrust to weight than could be achieved using other fuels. The fuel choices made this second stage a difficult development effort: the mixture was highly volatile and could not be readily handled. But the stage could produce an additional 90,000 pounds of thrust.

Saturn I was solely a research and development vehicle at first, making 10 flights between October 1961 and July 1965. The first four flights tested the rocket's first stage, but from the fifth launch onward, the second stage, Centaur, was active, and the later missions were used to place scientific payloads and Apollo test capsules into orbit. Altogether, the work on Saturn I was a straightforward, systematic research and development effort, composed of incremental improvements that advanced the technology along a well-understood path.

The next step in Saturn development came with the maturation of the Saturn IB, an upgraded version of the earlier vehicle, sporting more powerful engines but otherwise essentially the same machine. Its first stage could now generate 1.6 million pounds of thrust at launch, and coupled with its upgraded Centaur second stage, it could place 62,000-pound payloads into Earth orbit. On its first flight, on February 26, 1966, the Saturn IB tested the capability of both the rocket stack and the Apollo capsule in a suborbital flight. Two more flights followed in quick succession. Then there was a hiatus of more than a year before the January 22, 1968, launch of a Saturn IB with both an automated Apollo capsule and a lunar landing module aboard for orbital testing. The only astronaut-occupied flight of the Saturn IB took place on Apollo 7 between October 11 and 22, 1968, when Walter Schirra, Donn F. Eisele, and R. Walter Cunningham made 163 Earth orbits to test Apollo equipment.

Reaching the Moon, however, would require a still larger rocket, and the Saturn V, tested beginning in 1967, was the culmination of all the earlier booster development and test programs. Standing 363 feet tall,

The power of the Saturn V is in evidence as Apollo 11 lifts off from the Kennedy Space Center on July 16, 1969. This photograph captures the emotion and power of the mighty Moon rocket's launch, and it too has been reproduced in myriad ways since that time. Hundreds of thousands of Americans made it possible to reach the Moon, and this launch was one of the most watched events in human history. (NASA image no. 69PC-0421)

with three stages—the S-IC first stage, the S-II second stage, and the S-IVB third stage—this was the first vehicle that could take astronauts to the Moon and return them safely to Earth. All three stages used LOX as an oxidizer, but the first stage used RP-1 for fuel, while the second and third stages used LH2. The first stage generated 7.5 million pounds of thrust from five massive engines developed for the system. Those engines, known as F-1s, were some of the most significant engineering accomplishments of the Apollo program, requiring the development of new alloys and new construction techniques to withstand the extreme heat and shock of firing. The thunderous sound of the first static test of this stage, which took place in Huntsville, Alabama, on April 16, 1965, convinced many NASA engineers that perhaps John F. Kennedy's goal was actually within technological grasp. People gathered to view the test, many not even going into the Marshall Space Flight Center to witness it but instead climbing to the tops of nearby foothills, near where the Appalachian Mountain chain begins. They saw the light from the engines firing first, and then the thunder of sound waves reached them a split second later. The test could be better experienced as a feeling than as something seen, they recalled; the pounding of the engines thumped against the chest and rang in the ears. No one walked away from the experience without realizing that something wondrous had happened.

For others, the Saturn V test's success signaled the magic of rocket technology; some even characterized rocket development as a "black art," without rational principles. The second stage of the Saturn V had presented enormous challenges to NASA engineers and very nearly caused the lunar landing goal to be missed because of its complex fuels. It was always behind schedule and required constant attention and additional funding to ensure completion by the deadline for the lunar landing. At the same time, the first and third stages of the Saturn V development program moved forward relatively smoothly, and in the end the rocket worked well every time it was used.

Building the Saturn V was no small task. Its development and deployment required the orchestration of thousands of engineers, managers, and technicians and the interplay of numerous corporations, universities, and other organizations. As a small measure of the complexity of this task, the principal components of the Saturn V required the

With the hard-hatted workmen at lower right providing scale, this image demonstrates the massive size of the 363-foot-high Saturn V rocket as it rolls to Launch Complex 39A on September 8, 1969. (NASA image no. 69PC-0529)

efforts of the Boeing Company to create the first stage; North American Aviation for the S-II second stage; the Douglas Aircraft Corporation for the third stage, the S-IVB; the Rocketdyne Division of North American Aviation for the J-2 second-stage and F-1 first-stage engines; and IBM for the Saturn's instruments. These primary contractors, along with more than 250 subcontractors, provided millions of parts and components for the Saturn V launch vehicle, all meeting exacting specifications for performance and reliability. The total cost of the rocket's development was massive for the time, amounting to $9.3 billion. So huge was the overall Apollo endeavor that NASA's purchases of supplies and equipment, as well as major hardware such as the command and service module, lander, and booster, rose from roughly 44,000 specific purchases in 1960 to almost 300,000 by 1965.

Although ultimately successful and, in retrospect, a seemingly smooth process, rocket development during the 1960s was driven by trial and error, and sometimes the errors were spectacular. Early rockets tended to fail more often than the optimistic engineers building them were willing to admit. In reality, 13 percent of all NASA rocket launches conducted between 1957 and 1975 failed, a rate much like that of the larger rocketry effort (military, scientific, and commercial) during the same period. Without question, NASA's early spaceflight record could support accusations of failure as easily as it does images of accomplishment. For example, in 1958, all four of NASA's major launch attempts failed, and the next year, six of 14 attempts failed. The situation was little better in 1961, when nine of 24 NASA major launch attempts failed. The media could have roasted NASA for these troubles, but, interestingly, the space agency escaped such treatment in most cases. This may have been because of the Apollo program's national security dimension; Americans have been forgiving of military failures if they are convinced that eventual success will result. Indeed, at the start of the Corona satellite reconnaissance program in 1959 and 1960, 13 straight equipment failures occurred before the 14th, launched on August 10, 1960, at last went smoothly. No one had a high tolerance for failure in the first decade of spaceflight, but it proved character-building as NASA engineers worked to perfect the technology.

Of course, NASA got better over time, learning from its earlier mistakes. Between 1976 and 1999, the space agency had fewer than 4 percent of its launches fail, a relatively significant transformation from the beginning of the space age, and NASA has suffered very few launch failures since the 1970s. This reliability rate is the world's best: since the beginning of the space age, the rate has been less than 80 percent for all other nations that have undertaken space launches.

In NASA's early years, though, the agency found that even when its rocket boosters performed properly and lifted spacecraft beyond Earth, sometimes the satellites onboard failed in space. During the first five years of the space agency's existence, 36 percent—more than one-third—of the 102 missions reaching space eventually failed, and some of these failures were significant. For example, the first six Ranger probes, a series of robotic missions designed to produce close-up pictures of the Moon, failed to accomplish their task. Congress launched an investigation, but the incidents hardly tarnished NASA's reputation. The agency recovered and turned Ranger into a successful program with the last three probes in 1965–66, and spaceflight controllers gradually improved their work. NASA missions conducted between 1958 and 1970 suffered a 25 percent failure rate, but by the end of the 20th century, mission reliability, meaning that an effort achieved its goals, rose to an impressive 95 percent and has since remained there.

NASA's biggest issue with Saturn V's development for the Moon-landing program lay not in the hardware but in a clash of philosophies over development and testing. The Von Braun rocket team had made important technological contributions and enjoyed popular acclaim because of their conservative engineering practices, which took minutely incremental steps in testing and verification. The team tested each component of each system individually and then assembled them for a long series of ground tests. Then the team would launch each stage individually before assembling the entire system for another long series of flight tests. While this practice ensured thoroughness, it was both costly and time-consuming, and NASA had neither commodity to expend.

Among those who disagreed with the Von Braun approach was George E. Mueller, the head of NASA's Office of Manned Space Flight. Diminutive of stature but broad in personality and confident in his engineering sense, Mueller drew on his experience in the Air Force and the aerospace industry to head the Apollo program. Regardless, the program slipped as technical problems mounted and resolutions lagged. He found that the engineers wanted to test every individual component to the breaking point, then assemble subsystems and do the same with them. At the rate that they were proceeding, Mueller was certain, NASA would not make its appointment with the Moon "before this decade is out." To get back on track, and shadowed by the twin bugaboos of schedule and cost, Mueller advocated what he called the "all-up" concept, in which the entire Apollo-Saturn system was tested together in flight, without laborious preliminaries. A calculated gamble, the first Saturn V test launch took place on November 9, 1967, and included the complete Apollo-Saturn combination. A second test followed on April 4, 1968, and although it was only partially successful—the second stage shut off prematurely and the third stage, needed to launch the Apollo payload into lunar trajectory, failed—Mueller declared that the test program had been completed and that the next launch would have astronauts aboard. His bets paid off. Over 17 tests and 15 piloted launches, the Saturn booster family scored a 100 percent launch reliability rate.

NASA began its second intensive technical task—developing a spacecraft that could safely get astronauts into lunar orbit and back to Earth—in late 1961. Engineers came up with a complex system. It consisted of a three-person command module capable of sustaining human life for two weeks or more in either Earth orbit or a lunar trajectory; a service module holding oxygen, fuel, maneuvering rockets, fuel cells, and expendable and life-support equipment that could be jettisoned upon reentry to Earth's atmosphere; a retrorocket package attached to the service module, which would slow it down to prepare for reentry; and, finally, a launch escape system that could carry the crew to safety in the event of a rocket failure during launch and would be discarded upon achieving orbit. The teardrop-shaped command module had two hatches, one on the side for the crew's entry and exit at the beginning and end of the flight, and one in the nose, with a docking collar, for use in moving to and from the lunar landing vehicle.

Work on the Apollo spacecraft stretched from November 28, 1961, when the prime contract for its development was given to the North American Aviation Corporation, until October 22, 1968, when the last test flight took place. In the intervening years, NASA pursued various efforts to design, build, and test the spacecraft both on the ground and in suborbital and orbital flights. Not everything went well. In late 1965, Mueller stepped in and ordered his Apollo head, Major General Samuel C. Phillips, to investigate what was going on at North American Aviation's plant and determine why work on the Apollo spacecraft was lagging behind schedule while the budget ballooned. Phillips found that since the beginning of the spacecraft program, the capsule had "slipped more than six months. In addition, development of the first manned and the key environmental ground spacecraft have each slipped more than a year," he wrote in a letter to North American. "These slippages have occurred although schedule requirements have been revised a number of times, and seven articles, originally required for delivery by the end of 1965, have been eliminated."

The NASA project manager for the Apollo spacecraft, Joseph F. Shea, took this hard. He had been overseeing the spacecraft's design and construction for years with verve and style, but his insistent efforts did not yield the standard of excellence he wanted. Shea, who specialized in systems engineering and integration and took a holistic approach toward controlling every aspect of the project, began to browbeat his contractors, other NASA officials, and experts working on the system. This might have been counterproductive: one of his targets was the Von Braun rocket team at Marshall Space Flight Center, and his intrusion into decisions that the rocketeers believed were within their own purview led to flare-ups that had to be resolved by testy meetings between Von Braun and Houston's director, Robert Gilruth. Regardless, as Mueller later recalled, Shea "contributed a considerable amount of engineering innovation and project management skill." While those working for him enjoyed his eccentricities—especially his bad puns and decidedly unprofessional clothing choices—they cringed as he too often made himself a nuisance by moving into the construction site and sleeping on a cot during crucial times. By the end of 1966, Shea wrongly believed he had a spacecraft ready for human occupancy, notwithstanding the findings of Phillips's contractor review, and he really did not want to hear otherwise. Yet he reluctantly passed along NASA's concerns and demands for radical reforms to his industry counterpart at North American Aviation, and Mueller made his own, more forceful statement to the contractor's president, J. Leland Atwood: "I have been in this business long enough to understand quite well the difficulties and setbacks that occur and manifest themselves in . . . the development, building, and operation of sophisticated systems involving advanced technology . . . a good job has not been done. Based on what I see going on currently, I have absolutely no confidence that future commitments will be met."

Subsequent reorganizations in North American Aviation's management made some difference, perhaps, but a little more than a year after this scathing investigation uncovered serious flaws in the space-craft program, the most serious accident of the Apollo program nearly derailed the whole effort. On January 27, 1967, Apollo-Saturn (AS) 204, scheduled to be the first spaceflight with astronauts aboard the capsule, was at Launch Complex 34 at Kennedy Space Center in Florida, moving through ground simulations in what was called a "plugs out" test. The three astronauts slated to fly on this mission—Lieutenant Colonel Virgil I. (Gus) Grissom, a veteran of both the Mercury and Gemini programs; Lieutenant Colonel Edward H. White, the astronaut who had performed the first extravehicular activity, or spacewalk, during the Gemini IV flight in 1965; and Lieutenant Commander Roger B. Chaffee, an astronaut preparing for his first spaceflight—were aboard, running through a mock launch sequence.

Just before 6:31 p.m., after several hours of work, the voice of one of the three astronauts spoke to the controllers: "We have a fire in the cock-pit." The small video camera trained on the hatch window revealed to test engineers that flames were engulfing the capsule. Screams heard on the intercom told the engineers the astronauts inside were both asphyxiating and burning to death. Then all went silent just before the fire's pressure breached the command module's hull.

Tanked up to spaceflight pressure with pure oxygen, the spacecraft had been filled with flammable materials, including yards of fast-burning Velcro. All it had taken to set off the disaster was a spark. It was eventually determined that the spark had originated beneath Grissom's seat; he may have stood on it, rubbing the metal seat against faulty, exposed wiring underneath it, and when an arc erupted in the electrical system, the pure-oxygen environment set off the Velcro like an explosion. In a flash, flames engulfed the capsule, and the astronauts died within seconds.

It took the ground crew five minutes to open the hatch. When they did so, they found three bodies. White evidently had been trying to open the capsule's hatch from the inside. The other two astronauts had been trying to find a place safe from the inferno. According to the accident report:

> From the foregoing it has been determined that in all probability the Command Pilot [Grissom] left his couch to avoid the initial fire, the Senior [White] remained in his couch as planned

for emergency egress, attempting to open the hatch until his restraints burned through and the Pilot [Chaffee] remained in his couch to maintain communications until the hatch could be opened by the Senior Pilot as planned. With a slightly higher pressure inside the Command Module than outside, opening the inner hatch is impossible because of the resulting force on the hatch. Thus the inability of the pressure relief system to cope with pressure increase due to the fire made opening of the inner hatch impossible until after cabin rupture, and after rupture the intense and widespread fire together with rapidly increasing carbon monoxide concentrations further prevented egress.

Shock gripped NASA and the nation during the days that followed the Apollo 1 disaster. These were the first deaths of astronauts in a spacecraft. Earlier astronauts had died in aircraft accidents, but only these were, to date, directly attributable to the US space program. NASA Administrator James E. Webb told the media, "We've always known that something like this was going to happen soon or later . . . [but] who would have thought that the first tragedy would be on the ground?"

As the nation mourned, Webb went to President Lyndon B. Johnson and asked that NASA be allowed to handle the accident investigation and direct the recovery. He promised to be truthful in assessing fault and pledged to assign it to himself and NASA management if appropriate. The day after the fire, NASA appointed an eight-member investigation board, chaired by Floyd L. Thompson, longtime NASA official and director of the Langley Research Center. The board concluded in its final report in April 1967 that the fire had been caused by a short circuit in the electrical system that ignited an excessively large amount of combustible materials in the spacecraft and had been fed by the pure-oxygen atmosphere. The board also found that the fire could have been prevented but that poor design, shoddy construction, and improper attention to safety and redundancy had virtually mandated an accident.

The board also made several specific recommendations that led to major design and engineering modifications and to revisions in test planning and discipline, manufacturing processes and procedures, and quality control. Among these was the recommendation, which NASA readily accepted, to move to a less oxygen-rich environment in the Apollo spacecraft. Changes to the capsule followed quickly, and within a little more than a year, it was ready to test in flight.

This tragic story did not end with discovery of the technical problems that had caused the command module fire, despite NASA's best efforts to end the intense scrutiny at that point. Congress held several hearings in both houses during the spring of 1967, and agency officials were grilled at each one, especially after the release of the Thompson investigation team's report. Webb seemed evasive and defensive in these hearings, his normal verbosity and pomposity perhaps exaggerated by the very apparent pressure. The *New York Times*, which had been sharply critical of both Webb and the Apollo program, had a field day with this situation, once quipping that NASA stood for "Never a Straight Answer."

At a fundamental level, the Apollo 1 capsule fire demonstrated that NASA was far from the infallible agency that the public imagination had embraced. Numerous social scientists have undertaken studies to understand the process that people go through in grieving, analyzing, and making sense of a tragic accident. In a devastating accident such as the Apollo 1 fire, this multistage process was wide and involved virtually all of American society. During the first stage, in the immediate aftermath of the accident, few blamed any particular person or group. Most sought instead to understand what had happened, and thereafter the investigation followed a step-by-step procedure: assessing responsibility, making alterations, and recovering from the tragedy over several months. Indeed, many people lauded the heroism of the crew and emphasized the inherent riskiness of spaceflight. Legendary NASA Flight Controller Gene Kranz gave a speech to his team after the accident that pointed to how the brave men who lost their lives in the capsule fire would be memorialized by everyone else redoubling their efforts:

From this day forward, Flight Control will be known by two words: *Tough* and *Competent. Tough* means we are forever accountable for what we do or what we fail to do. We will never again compromise our responsibilities. . . . *Competent* means we will never take anything for granted. . . . Mission Control will be perfect. When you leave this meeting today you will go to your office and the first thing you will do there is to write *Tough* and *Competent* on your blackboards. It will never be erased. Each day when you enter the room, these words will remind you of the price paid by Grissom, White, and Chaffee. These words are the price of admission to the ranks of Mission Control.

NASA also stressed that the US goal of reaching the Moon by the end of the decade would not be set back by the tragedy and that Grissom, White, and Chaffee had not died in vain; indeed, their deaths would strengthen American resolve to accomplish the job for which the three had made the ultimate sacrifice.

Several political cartoons appearing in major dailies around the United States are representative of public opinion in the immediate aftermath of the Apollo 1 fire. Most focused on the sacrifices of the astronauts. One, published the day after the fire, shows the crew locked arm in arm, walking into the distant cosmos. The caption reads, "And now they belong to the Heavens." Another graphically illustrated the risks involved in spaceflight and suggested that the United States must not curb its efforts because of the tragedy: a spacesuit-clad skeleton stands before a burning Apollo spacecraft on the launch pad, holding in its arms a Mercury and a Gemini capsule. The caption reads, "I thought you knew; I've been aboard on every flight."

The second and longest stage of the process was the official NASA inquiry, undertaken by Thompson's board and meant to establish what had happened and why. Periodic reports of progress in the investigation emerged, but in general public attention to the accident waned until the investigation was over and the final report released on April 10, 1967, kicking off the third stage in the recovery process. Reaction was immediate. The *Washington Evening Bulletin*'s opinion was typical in its reporting the day after the report's release: "Although the Board did not fix the precise cause of the fire, and named no persons or group as being responsible, it laid a heavy burden of negligence and poor workmanship on many. . . . That inquiry in Congress will go ahead. Perhaps it will not be content with blame so generally spread."

Several particularly telling political cartoons published in the spring of 1967 demonstrated emerging anger at NASA and its contractors for allowing the accident to take place. One, from the *Evening Star* in Washington, DC, depicts a burning Apollo capsule labeled "Apollo Report" and captioned "In a Word—Carelessness." Another shows the

Apollo/Saturn spacecraft on Pad 34, with tools labeled "safety precautions" and "quality control" strewn in the foreground. The caption reads, "Washington slept here." Yet another shows a member of the Thompson board standing before the graves of the three dead astronauts, saying, "It's official . . . you died from a case of lingering carelessness." In one of the most biting indictments, a drawing in Philadelphia's *Evening Bulletin* on April 15, 1967, depicts space contractors and members of the congressional oversight committee, with the caption "Carelessness, not negligence, caused the Apollo fire." Above the illustration is a statement that reads:

DICTIONARY:

Care.less.ness, n.—Syn.—Negligence.

Neg.li.gence, n.—Syn.—Carelessness.

Yet a general statement of institutional carelessness, whether legitimate or not, did not suffice to explain the Apollo 1 fire. Early in the congressional review of the fire, Webb had said that NASA "would accept our part of the blame and we are prepared to accept your [congressional] judgment as to what it is." Congress found this response unacceptable, consistently asserting that individuals are morally responsible for any inauspicious event and that if NASA as a whole was to be faulted, some of its officials had to share in that fault. Senator Walter Mondale (D-MN) condemned the agency for its "evasiveness . . . lack of candor . . . patronizing attitude toward Congress . . . refusal to respond fully and forthrightly to legitimate Congressional inquiries, and . . . solicitous concern for corporate sensitivities at a time of national tragedy." Although Webb attempted to focus blame at the organizational level, individuals were publicly named as culpable over the course of the congressional investigation. The first person to be faulted for the accident was Joseph F. Shea, Apollo program manager at Houston's Manned Spacecraft Center (renamed the Lyndon B. Johnson Space Center in 1973). Rocco Petrone, a director of the Marshall Space Flight Center in the 1970s, had been sitting next to astronaut Deke Slayton in the blockhouse at the Kennedy Space Center when the fire broke out. Petrone blamed Shea, as the Apollo spacecraft manager, for the errors that had led to the accident, supposedly telling him, "You are a menace and you are to blame for the fire. When you die, I will come and piss on your grave."

No one took the Apollo accident more personally than Shea and his counterpart at North American Aviation, Harrison Storms. Both were reassigned afterward, Shea in no small part because of the psychological toll the deaths of the astronauts took on him. He self-medicated with alcohol and barbiturates. Within a few weeks, Chris Kraft, senior flight director in charge of Mission Control, confided to colleagues in Houston that Shea's erratic behavior was hampering the recovery from the fire. He recalled of one meeting on the accident: "Joe Shea got up and started calmly with a report on the state of the investigation. But within a minute, he was rambling, and in another thirty seconds, he was incoherent. I looked at him and saw my father, in the grip of dementia praecox. It was horrifying and fascinating at the same time." Webb, too, worried about Shea, and asked him to come to Washington to help the Apollo program from headquarters, promising Shea that he would work only for Webb. He did not tell Shea that no one would be working for *him*. Putting Shea in a corner might have been the best thing both for him and for the Apollo program, but Shea disliked the move.

After being "kicked upstairs," Shea resigned from NASA on July 24, 1967, to become a vice president of the Raytheon Corporation. He served as a consultant to NASA in later years but never again worked directly for the agency. Although his replacement was not announced as a punishment, the public interpreted it as such, and had it not been for the accident, there is no reason to believe that Shea would have been replaced just as the program was nearing flight stage.

The Apollo fire had a remarkable effect on Shea throughout the remainder of his life. I witnessed that impact personally in April 1993, when NASA Administrator Daniel S. Goldin asked Shea to chair a review board concerning the redesign of what became the International Space Station. The space station effort at the time was over budget and behind schedule, and the newly established Clinton administration had directed NASA to either restructure the program or see it canceled. At a public meeting in the NASA headquarters auditorium, which I attended, Shea gave preliminary results of his board's findings. His talk started out fine, but then he launched into a two-hour monologue, with the 1967 Apollo fire as its centerpiece. He rambled about the horror of that experience some 26 years earlier, its relationship to the space station situation, the decline of NASA as a technical agency, and the way in which the space agency cavalierly, in his opinion, had placed astronauts' lives on the line. Then Shea wheeled back to an incoherent defense of his management of the Apollo spacecraft and how he was not at fault for the capsule fire. Finally Goldin stopped him and led him off the stage. I came away from the presentation saddened by Shea's condition, knowing that it did not represent his professional life and understanding more fully that the trauma of the accident had affected everyone associated with it. A day later, Goldin relieved Shea of his responsibility for the space station review board but, out of deference to his stature as a legendary NASA official, kept him on as an informal advisor.

Following Shea's 1967 reassignment, attention turned to establishing fault with NASA's primary Apollo spacecraft contractor, North American Aviation. The agency had been critical of workmanship and quality control at North American for years, and a long paper trail documented these concerns. They fixed attention on Harrison Storms, president of North American's Space and Information Division; William Snelling, his executive vice president; and two other executives in the division. NASA forced J. Leland Atwood, chief executive officer for North American, to replace all four in April 1967. With these personnel changes, presumably those responsible for the fire had been punished, the American public had answers, and the nation and the Apollo program could move on.

Storms and other officials at North American were infuriated that he had to be sacrificed to get the company back into NASA's good graces and recover from the accident, and he never forgave NASA for having to take the fall. He stridently blamed NASA for using a 100 percent oxygen environment in the capsule, something that had intensified the fire, over North American's objections. Storms was also resentful that, although he had expressed his belief that the risk of fire was too great, his company and Atwood quietly accepted NASA's condemnation because of the business the space agency gave them. Storms's arguments that he and North American were innocent scapegoats would rile NASA long after the end of the Apollo program; in the early 1990s, Storms worked with Hollywood screenwriter Mike Gray to tell his side of the story in *Angle of Attack: Harrison Storms and the Race to the Moon*. Based almost exclusively on the recollections of Storms and others at North American and published in 1992, the book serves as a counterweight to NASA's official story of malfeasance by contractors. Yet it was so flawed by errors of fact great and small, one-sided pleading, name-calling, and a glossing over of corporate culpability in favor of damning NASA that few took it seriously.

Although the ordeal was personally taxing to all who played any role in the Apollo fire, Webb's strategy deflected much of the resulting critique from both NASA as an agency and from the Johnson administration. Although Webb was personally tarred by the disaster and certain members of the congressional subcommittee asked for his resignation, the space agency's image and popular support were largely undamaged. After the hearings ended in May 1967, Congress remained supportive of NASA's continued efforts to place an American on the Moon by the end of the decade.

By 1968, questioning of NASA's ability seems to have abated, and it was almost entirely swept away by the successful lunar landing missions between 1969 and 1972. The change began with the October 1968 flight of Apollo 7. Though few remember the Apollo 7 mission as holding great significance, it was an enormous confidence-builder not only for NASA and those working in the space program but also for the public at large. As an Earth-orbital shakedown, the Apollo 7 mission proved the spaceworthiness of the Apollo spacecraft, and it did much more by reenergizing the widespread belief that the Moon landings were possible and that NASA could accomplish them.

A sampling of political cartoons from major dailies around the country in late October finds Apollo 7 the subject of sustained praise. One shows a beaming Uncle Sam in a Greek toga reclining on clouds with the Moon in the background. He holds an Apollo/Saturn stack in his arms, as if planning to toss it like a javelin, and sports a gold medal around his neck. The caption reads, "A Gold Medal in the Lunar Olympics." Another cartoon of the time shows a suspension bridge being built from Earth to the Moon, on its roadway the statement "Apollo 7 Success." The caption reads, "Almost Ready for the Ribbon-Cutting." A third editorial cartoon, titled "Another Winner," shows dice coming up "Apollo Seven" emerging from a shaker labeled "Space Risks." The *Christian Science Monitor* showed Apollo 7 splashing down in the ocean silhouetted against a full Moon in the background with the caption "Splashdown on the Way Up."

The Saturn launch vehicle and the spacecraft were difficult technological challenges, yet it was the third part of the hardware for the Moon landing, the lunar module (LM), that posed the most serious problem. Begun a year later than the other vehicles, it was required to safely convey a crew to a soft landing on the surface of another world and later return them to a lunar orbit where the crew could be reunited with the command module for the trip back to Earth. None of these things had been done before, so it is perhaps understandable that the LM was consistently behind schedule and over budget.

Much of the problem turned on the demands of devising two separate spacecraft components—one for descent to the Moon and one for ascent back to the command module—that maneuvered only outside an atmosphere. Both engines had to work perfectly, or it was quite possible that the astronauts would not be able to return home. Guidance, maneuverability, and spacecraft control also caused no end of headaches. The landing structure likewise presented problems; it had to be light, sturdy, and shock-resistant. In November 1962, Grumman Aerospace signed a contract with NASA to produce the LM, and work on it began in earnest. An ungainly vehicle emerged that two astronauts could fly only while standing. The two-stage craft featured a lower landing stage, which included the landing gear, the descent rocket engine, and lunar surface experiments, and an upper ascent stage that consisted of a pressurized crew compartment, equipment and supply stowage, and a solid rocket ascent engine. After various engineering problems, a vehicle emerged that was finally declared flight-ready in January 1968. With difficulty, NASA orbited the first LM on a Saturn V test launch later the same month and judged it ready for operation. Thomas J. Kelly, Grumman's chief designer of the LM, later recalled the difficult task:

> The command module was totally dominated by the need
> to reenter the Earth's atmosphere, so it had to be dense and
> aerodynamically streamlined and all that, whereas the lunar
> module didn't want any of that. It wanted to be able to land on the
> Moon and operate in an unrestricted environment in space and
> on the lunar surface. It ultimately resulted in a spindly, gangly-
> looking, very lightweight vehicle that was just the opposite of all
> the attributes of the command module. If you tried to do that
> all in one vehicle, it would be a real problem. I don't know how you
> would have done it. But this way, with this mission approach, it
> was very neatly divided in two halves.

The lunar module proved its mettle during the first two piloted test flights and the landings of the first two Apollo missions on the Moon's surface. During Apollo 9 (March 3 to 13, 1969), the crew tested the LM in Earth orbit; on Apollo 10 (May 18 to 26, 1969), the LM performed well in lunar orbit, getting as close as 8.5 nautical miles to the surface. It also performed well in the landing of Apollo 11 on the Moon on July 20, 1969, when Neil Armstrong took the controls during descent to the surface to avoid a rock-strewn landscape and safely land in the Sea of Tranquility. Although he nearly ran out of fuel during this maneuver, Armstrong also inadvertently proved the capability of the Apollo landing craft. Astronaut Pete Conrad did the same in November 1969 when he landed the Apollo 12 craft within approximately 160 meters of the Surveyor III soft lander, which had been sitting on the lunar surface since 1966. Conrad and fellow astronaut Alan Bean walked over and retrieved several pieces from the spacecraft, including its television camera and some associated electrical cables, the sample scoop, and two pieces of generic aluminum tubing. As an exercise in precision landing, this mission demonstrated beyond all doubt that the LM, when coupled with a skilled pilot, was an impressive vehicle.

But the most impressive use of the LM came during the nearly disastrous Apollo 13 mission in April 1970. By that time, the Moon lander had already performed well and few doubted its capabilities, but by serving as a lifeboat for the crew, the LM solidified perceptions of NASA's technological virtuosity. On April 13, 56 hours into the flight, an oxygen tank in the Apollo service module ruptured, damaging several of the power, electrical, and life-support systems. People throughout the world watched and waited and hoped as NASA personnel on the ground and the crew, well on their way to the Moon and with no way to return until they went around it, worked together to find a safe way home. NASA engineers quickly determined that the Apollo capsule had insufficient air, water, and electricity to sustain the three astronauts until they could return to Earth, but they also found that the LM—a self-contained spacecraft unaffected by the accident—could provide austere life support for the return trip. It was a close call, but the crew returned safely on April 17. One might best refer to Apollo 13 as a successful failure, in no small part because of the life support provided by the LM.

Public reaction during the close call was spirited. The *Baltimore Sun* wrote: "The rest of us can only wait some desperate hour to hour, trusting that superlative skill, universal hope and fervent prayers will prevail over cruel chance." Howard Simons of the *Washington Post* commented that America's "sanguine attitude about manned space flight" had been shattered by the problems of Apollo 13. "As the first words came that the three astronauts were in peril," he wrote, "earthbound fears began to race along with the Moonbound craft. The astronauts seemed remarkably calm, most other persons desperate." That sense of desperation permeated the United States during the evening of April 13; people everywhere, Simons wrote, "whether lowly or mighty, paused to participate in the first life-and-death drama in deep space."

As the drama played out in space, the American press commented on the meaning of the Apollo 13 accident for the larger aspects of national character. The *Washington Post* editorialized:

> Lindbergh gave up a continent, as the astronauts gave up a planet, because he had measured the risks and found them reasonable, for all the possibility of disaster lurking around the corner—it does so every day for men and nations. So if we are going to apply the glories, we must confront the dangers along the way. All we can do now, as that unlikely looking craft limps back from the moon, is to hope that the men who ride in it and the men in Houston who guide it can find what Lindbergh found. "Somewhere in an unknown recess of my mind," he wrote, "I've discovered that my ability rises and falls with the essential problems that confront me. What I can do depends largely on what I have to do to stay alive and on course."

The near-disaster of Apollo 13 left the nation with several legacies. First, it reemphasized the lesson, learned by the public after the Apollo 1 fire, that spaceflight had never been and was still not a routine activity. It was inherently risky, and those who participated in it, either directly or vicariously, had to be prepared for the loss of life. Second, Americans have long exemplified a great resiliency when dealing with adversity, and the near-tragedy of Apollo 13 reunited the country and reinforced its commitment to space exploration as a national priority.

Third, and perhaps most important, the successful recovery of the astronauts solidified in the popular mind NASA's technological competence and the place of the Moon-landing effort as representative of an America that could do anything it set out to accomplish. It is ironic that such an episode in the Apollo program has been transformed into one of the greatest achievements of the whole endeavor. Flight Director Eugene Kranz has been credited with saying, "Failure is not an option" during the desperate hours in Houston when NASA engineers were working to bring the crew home alive. Although the line was dreamed up later by a Hollywood script-writer and Kranz, in retrospect, only wished he had been so eloquent, both he and the flight team deserve high marks for perseverance, dedication, and an unshakable belief that they could bring the crew back safely—to say nothing of the professionalism of the crew in space. The fact that Apollo 13 is now viewed as one of NASA's shining moments says much about the ability of humanity to recast historical events into meaningful morality plays and assign them a central place in our memories.

Apollo 13's place at the center of NASA's vaunted reputation was extended by the powerful 1995 feature film *Apollo 13*, which appeared in theaters throughout the world and was based on astronaut Jim Lovell's bestselling autobiography, *Lost Moon*. Director Ron Howard set out to make a riveting film whose ending everyone watching already knew, in which none of the characters were villains, in which none of the sex, language, and violence common to big-budget films was present, and in which NASA's technocratic nerds were held up for praise rather than ridicule. The film remains tense despite the obvious challenges, and the nerds become both heroes of the film and personifications of the American spirit. More important, they demonstrate repeatedly the nature of both NASA's commitment and capabilities. Interestingly, the lunar module is the unsung hero of this film, serving the astronauts well until it is discarded for reentry to Earth.

The fourth major technological issue that had to be overcome involved guidance and control, navigation, and computing. Many mundane tasks had to be worked out, such as the establishment of a worldwide system of voice and data communications for the Apollo astronauts in their various vehicles operating in space. NASA engineers fully understood what was required, and they set about establishing ground stations and building the transceivers needed for the space vehicles. The Deep Space Network, as it became known, tracked and communicated guidance and control data from several stations. Goldstone, the first US station designated for this purpose, became operational in 1958 in California's Mojave Desert and would support the missions to the Moon. Stations in Australia, South Africa, and Spain were added later. Each facility had at least four large dish antennas used to transmit and receive data from operational spacecraft. They took thousands of measurements of the Apollo crafts' distance and velocity and helped chart the trajectory of the missions, all tasks critical to the success of the Apollo program.

The network's staff plotted the course of spacecraft and determined their movements by measuring the Doppler shift of their radio transmissions. Recall that radio waves are a type of electromagnetic radiation and travel at the speed of light. All spacecraft are designed to transmit at certain radio frequencies. On Earth, the received signal is slightly shifted to a lower frequency because the spacecraft is moving away, in a phenomenon similar to the one that causes light from stars and other galaxies to be red-shifted. Measuring the amount of Doppler shift enabled Apollo engineers to determine the rate at which the spacecraft was receding. The range, or distance, of a spacecraft also had to be determined, which could be accomplished via radio transmissions as well. The amount of time that a signal needs to travel between Earth, the probe, and back can be used to determine the craft's distance because the speed of light is precisely known. Engineers also knew the time required for the signal to be processed on the spacecraft, gleaned by testing copies of the craft's radio equipment. Earth's movements between the time of transmission and when the signal was received had to be known as well, as did the angular position of the spacecraft, a measure of its location as seen from Earth. Angular position could be crudely measured by using information from an Earth-based antenna, and more precisely determined by using transmissions received simultaneously at two different Earth-based stations. Along with range and velocity, this data allowed engineers to calculate the craft's position and movement in three dimensions.

A truly revolutionary breakthrough in guidance and control came via development of the Apollo onboard computer. NASA engineers realized that they would have to invest in research and development for miniaturized electronic components, and among the other results were the first computers that were small enough to fit into the Apollo command module. Until that time, computers routinely filled large rooms and required specialized air, handling, and power systems and a dedicated staff to keep the finicky equipment operating. Accessing these computers was a complex process of developing programs on punch cards, and results were retrieved more than a day later, printed out on reams upon reams of paper. Adapting such a computer for use on a small spacecraft proved to be a great challenge, but NASA pulled it off.

Much of the work was done at MIT's Draper Laboratory, where engineers such as Eldon Hall worked to first design and then miniaturize the Apollo guidance computer, which had limited capability that nonetheless far exceeded anything previously seen. The Display and Keyboard (DSKY) was developed as a general-purpose system that allowed for considerable flexibility in presenting information. Each command module carried two identical DSKYs, each connected to the spacecraft's guidance computer. The lunar module also had a single DSKY connected to its computer. Through them, astronauts communicated with the craft's computers to perform critical navigation, rendezvous, docking, landing, ascent, and flight management tasks throughout a mission. An astronaut could load coordinates into the computer by keying in numeric codes in a simple and intuitive manner, and this ease of use proved critical both for the Apollo program and for the technology's migration to commercial uses in the 1970s. The onboard computer proved its worth repeatedly during the Apollo program, the astronauts even dubbing it the "fourth crew member," since they used it in almost every phase of a flight. Most important, it pointed up the symbiosis that had developed between the technology of the Moon program and the astronauts who flew on the spacecraft.

None of these technologies existed when the Apollo program began, but all were within ready grasp. NASA engineers employed a "program management" concept that centralized authority over design, engineering, procurement, testing, construction, manufacturing, spare parts, logistics, training, and operations for all the technologies necessary to reach the Moon. The management of the program was recognized as critical to Apollo's success in November 1968, when *Science* magazine, the publication of the American Association for the Advancement of Science, observed:

In terms of numbers of dollars or of men, NASA has not been our largest national undertaking, but in terms of complexity, rate of

growth, and technological sophistication it has been unique. . . .
It may turn out that [the space program's] most valuable spin-
off of all will be human rather than technological: better
knowledge of how to plan, coordinate, and monitor the
multitudinous and varied activities of the organizations required
to accomplish great social undertakings.

Effectively managing the complex organization and systemic structures necessary for the successful completion of such a multifaceted task was a critical component of NASA's technological virtuosity, which played to the story of American achievement that still dominates the memory of the Apollo program. The technologies necessary to go to the Moon, while sophisticated and impressive, had been largely within the grasp of the United States even at the time of the 1961 decision. More difficult, and perhaps more impressive, was ensuring that those technological skills were properly managed and used.

How do we account for the rise of NASA's technological virtuosity in the 1960s and its continuation to the present? Perhaps it is the high pedestal upon which science and technology have long been perched in American society, a position that predisposed the public to embrace Apollo as a repre-sentation of American greatness. Many of the great accomplishments of the US government in the 20th century have involved science and tech, includ-ing the construction of the Panama Canal, construction of Boulder Dam and related efforts that made possible the hydraulic culture of the American West, the Manhattan Project, the Salk vaccine, the interstate highway system, and Project Apollo. Through Apollo, two American presidents came to appre-ciate the power of science and technology to increase confidence in the US government both abroad and at home. Indeed, at a fundamental level both John F. Kennedy and Lyndon B. Johnson consciously used Apollo as a symbol of national excellence to further their objectives of enhancing the prestige of the United States throughout the 1960s. At a fundamental level, too, Apollo fed the deep-seated affection that Americans feel for all things technological, establishing in concrete form a long-held belief that society's problems could be conquered and the world made a more perfect place through the harness-ing of human intelligence and machines of great complexity.

Nurtured in the political climate of progressivism at the turn of the century, with its emphasis on professionalism and expertise, Americans of the Apollo era had been taught that scientific and technological knowledge could solve almost any problem. Social reformer Edward Scribner Ames, for instance, reflected just after World War I that the best way to prevent such massive destruction again was to pursue science and tech with increased enthusiasm. He believed scientists had saved the Allied powers from conquest, and that in the future technology would provide American society all types of new conveniences, medical capabilities, and improvements to daily life.

The immediate postwar era saw the application of wartime mobilization models for science to peacetime problems. In 1952, Edward Everett Hazlett, a lifelong friend of Dwight D. Eisenhower, wrote to the presidential candidate about declaring a "War on Untimely Death." He suggested that a widespread government effort to "smash the atoms" of disease "seems no more likely to fail than did that on the atom. It has, in addition, the spiritual advantage of being a campaign to save life and not to take it." Such faith in science and technology motivated all manner of activities in the 20 years after World War II, and government officials yielded to the authority of experts with something akin, according to Harvard University President James B. Conant, to "the old religious phenomenon of conversion."

These perspectives were also abundant in the Kennedy administration. David Halberstam shrewdly observed that "if there was anything that bound the men [of the administration], their followers, and their subordinates together, it was the belief that sheer intelligence and rationality could answer and solve anything." This philosophical belief translated into an ever-increasing commitment to science and tech to resolve problems and point the direction for the future. The approach infused both issues of public policy and international relations, and the space program and the technologically driven war in Vietnam were two direct results.

By the time of the Tet Offensive in Vietnam in 1968, however, it was clear to many Americans that science and technology did not hold the answers that had been promised. All the bomb tonnage, all the modern military equipment, and all the supposed expertise had not defeated the simply clad and armed North Vietnamese. Nor had tech been capable of eradicating disease, ending world hunger, resolving racial strife, stamping out poverty, fostering human equality, enhancing the level of education, or settling a host of energy and ecological issues. In too many instances, or so it seemed, science and tech were viewed as fundamental parts of the problems, not as the solutions that they had once seemed. Yet Apollo partially staved off this general malaise about American capability until new criticisms of technocracy and expertise came to the fore in the latter part of the 20th century.

In the early decades of the 21st century, while there may be more caution on the part of individuals in adopting innovative technologies—witness the current debate over genetically modified foods and the threats to individual privacy posed by massive telecommunications methods/ corporations—American society remains enthusiastic about tech of all types. At a basic level, we have accepted what theoretical physicist Ralph E. Lapp once characterized as the ordination of technical experts as a "New Priesthood," deferring to them as elites who are better pre-pared to give answers to tough questions than anyone else. In 1965, Lapp warned, "Like any other group in our society, science has its full share of personalities—wide-gauge and narrow-track minds, sages and scoun-drels, trail-blazers and path-followers, altruists and connivers. . . . To say that science seeks the truth does not endow scientists as a group with special wisdom of what is good for society." Nonetheless Apollo remains so positively viewed an example of American technological prowess that it has been rendered nearly mythical.

For the generation of Americans who grew up in the 1960s, watching NASA astronauts fly into space, beginning with 15-minute suborbital tra-jectories and culminating with six landings on the Moon, Project Apollo signaled in a very public manner how well the nation could do when it set its mind to it. Television coverage of real space adventures was long and intense, the stakes high, and the risks to life enormous. There were moments of both great danger and high anxiety. In the whole decade, however, NASA lost not a single astronaut during a space flight, and Project Apollo proved itself a triumph of management as it met enor-mously difficult systems engineering, technological, and organizational integration requirements.

The Apollo successes have remained in the forefront of American con-sciousness to the present day, and the Moon-landing program is still some-times used to exemplify the best Americans can bring to any challenge, deployed to support the nation's sense of its own greatness. Actor Carroll O'Connor perhaps said it best in a 1971 episode of *All in the Family,* the show in which he portrayed the character of Archie Bunker, the bigoted working-class American whose perspectives were more common in our society than many observers were comfortable admitting. Archie tells a visitor to his house that he has "a genuine facsimile of the Apollo 14 insignia. That's the thing that sets the US of A apart from . . . all them other losers." In very specific terms, Archie Bunker encapsulated for many what set the United States apart from other nations: success in spaceflight.

A later pop-culture reference serves as evidence of the durable sense of accomplishment granted to America by the Moon landings. In a 2000 episode of the critically acclaimed television situation comedy *Sports Night*, centered on a team that produces a nightly cable sports broadcast, the fictional show's executive producer, Isaac Jaffee, played by Robert Guillaume, is recovering from a stroke and has disengaged from the daily hubbub of work. His producer, Dana Whitaker, played by Felicity Huffman, keeps interrupting him as he reads a magazine about space exploration. Isaac tells her, "They're talking about bioengineering animals and terraforming Mars. When I started reporting Gemini missions, just watching a Titan rocket lift off was a sight to see." Isaac affirms his basic faith in NASA to carry out any task in space exploration: "You put an X anyplace in the solar system," he says, "and the engineers at NASA can land a spacecraft on it."

The aura of technological virtuosity remains to this day, supporting an emphasis on national greatness and offering solace in the face of other setbacks. At the same time, for a subset of Americans, the Moon landings raise the specter of technocracy, of human freedom under the sway of powerful systems that may well be out of control. They see the overreliance of society on the technology that makes life easier as misplaced, and Apollo as the tipping point for the electronics revolution of the 1970s and the host of transformations that followed. Did Apollo lead to the world's people walking around while staring at their smartphones? The line is not straight, but the technology that was developed through Apollo did help, in part, to make the modern world what it is, and changed forever the manner in which humans live their lives.

4

Heroes in a Vacuum

"I still get a real hard-to-define feeling down inside when the flag goes by," said astronaut John Glenn to a joint session of Congress after his first flight into space in 1962. His comments epitomize the dominant recollection of the race to the Moon, one that celebrates pride in America and success in all its national endeavors. Although NASA hoped, from the announcement of the Mercury Seven in 1959 to the present, to shape its astronauts into national archetypes, few could have predicted in the early days of Apollo just how thoroughly the astronauts would come to embody their era and the accomplishments and values of the nation. They became the face of the space program, and to some extent the face of America.

Both NASA officials and the astronauts themselves molded and controlled their public images every bit as carefully as those of movie idols or rock stars, presenting themselves as heroes in the vacuum of space, to be sure, but also in the vacuum of American culture as they engendered goodwill around the globe. Their bravery touched emotions deeply seated in the human experience. Apollo astronauts could pilot spacecraft traveling at more than 25,000 miles per hour to the distant Moon, 240,000 miles beyond Earth, with virtually no room for any human error or mechanical malfunction, in journeys echoing Charles Lindbergh's "lone eagle" crossing of the Atlantic Ocean in 1927. Many observers commented on the skills and professionalism of this unique, daring, and exceptionally able group of individuals, the best the nation had to offer. Facing personal danger, they fit the myth of frontier heroes, whose grit had long provided the substance of Hollywood films; as former military test pilots in an era when service was held in high regard, they recalled the sacrifices required for the Allies to emerge victorious from World War II. Their personal exploits celebrated danger, ritual, and service to the nation. At the same time, the astronauts, about whom the public clamored for personal details, came to personify the "all-American" boys whose mythical lives

Lapel pins celebrating the astronauts became popular while the Apollo missions to the Moon were underway, and they remain an important collectible from auction houses and antique stores. (Photograph by Eric Long, NASM image no. 2005_36189.12)

populated family-oriented television programs during the 1950s and '60s. The astronauts' public personas emphasized bravery, patriotism, and religious faith, all complemented by loving wives and children.

Early in the space agency's existence, it contracted with *Life* magazine to cover the Mercury astronauts. *Life* paid NASA for the privilege, and the agency was allowed to review stories and photos before they were published. But the cost was worthwhile: the astronaut stories appeared at a time when NASA desperately needed to inspire public faith in its ability to carry out the nation's goals in space. Rockets might explode, but the astronauts shined. How could anyone distrust a government agency epitomized by such people? The trust granted to the astronauts themselves was easily extended, in the public mind, to NASA and to the larger government. As one of the *Life* reporters summarized: "[We] treated the men and their families with kid gloves. So did most of the rest of the press. These guys were heroes; most of them were very smooth, canny operators with all of the press. They felt that they had to live up to a public image of good clean all-American guys, and NASA knocked itself out to preserve that image."

Despite the wishes of the NASA leadership, the astronauts' fame quickly grew out of all proportion to their activities. Perhaps it was inevitable that they were destined for premature adulation, what with the enormous public curiosity about them, the risks they would take in spaceflight, and their exotic training activities. In a feedback loop, commercial competition for publicity and pressure for political prestige in the space race also whetted the insatiable public appetite for this new kind of celebrity.

From the beginning, astronauts emerged as noble champions who would carry the nation's manifest destiny beyond its shores and into space. James Reston of the *New York Times* exalted the astronaut corps when it was first unveiled in 1959, saying he felt profoundly moved by the press conference introducing the Mercury Seven astronauts and that even reading the transcript of it made one's heart beat a little faster and step a little livelier. "What made them so exciting," he wrote, "was not that they said anything new but that they said all the old things with such fierce convictions. . . . They spoke of 'duty' and 'faith' and 'country' like Walt Whitman's pioneers. . . . This is a pretty cynical town, but nobody went away from these young men scoffing at their courage and idealism."

These statements of values seem to have been totally in character for what was a remarkably homogeneous group. The astronauts all embraced a traditional lifestyle that reflected the mainstream values of American culture, playing up feelings about the role of family members in their lives and the effect of the astronaut career on their spouses and children. In a study analyzing several Apollo-era astronauts' autobiographies, sociologist Phyllis Johnson found that the public nature of what they did meant that their family and work lives were essentially inseparable and often took a toll on those involved in the relationship:

The data on these early astronauts need to be interpreted in light of the work-family views of the time: men were expected to keep their work and family lives compartmentalized. Their family life was not supposed to interfere with work life, but it was acceptable for work life to overlap into their family time. In high-level professions, such as astronauts, the wife's support of his career was important; rather than "my" career, it became "our" career. The interaction between work and family is an important aspect of astronaut morale and performance, which has been neglected by researchers.

The media, reflecting the desires of the American public, consumed the astronauts and their families at every opportunity, prompting NASA to construct boundaries that both protected the astronauts and projected specific images that reinforced the already dominant traditional struc-ture of American society. NASA, for obvious reasons, wanted to portray an image of happily married astronauts, with no extramarital scandals or divorces. Gordon Cooper, one of the original Mercury astronauts, recalled that NASA sought to hide such things as marital unhappiness for fear it might "lead to a pilot making a wrong decision that might cost lives—his own and others." Safety concerns might have been part of it, but the agency's leaders also wanted to protect the squeaky-clean image of the astronaut in no small measure because of their sense that any con-troversy could jeopardize public support for the lunar landing program.

As a result, the astronauts sometimes felt NASA's generally benevolent public-affairs flacks breathing down their necks. In turn, the astronauts sometimes caused the agency's officials considerable grief, generally over the rules the astronauts were expected to follow and the lack of well-understood guidelines for their ethical conduct. For example, when the Space Task Group moved to Houston in 1962, several local developers offered the astronauts free houses. The offer caused a furor that reached the White House and prompted the involvement of Vice President Lyndon B. Johnson. In this case, the head of the Manned Spacecraft Center, Robert R. Gilruth, had to disallow an outright gift to the astro-nauts. Gilruth's boys also got into trouble over what they could and could not do to make additional money beyond their paychecks. NASA's facilita-tion of the Mercury Seven's sale of their personal stories to *Life* magazine earlier in the decade had already stirred up a furor. NASA policies were changed thereafter, but in 1963 Forrest Moore, a Texas oil executive, com-plained to LBJ that the new astronauts were seeking to do essentially the same thing. NASA leaders had to intervene and explain that any deals for such stories would be worked through the NASA general counsel and could take place only in a completely open and legal manner. More than once, Gilruth had to defend the astronauts to NASA leadership, as when they accepted tickets to see the Houston Astros' season-opener baseball game in the new Astrodome in 1965. While he told his superiors that he saw no reason why the astronauts should not enjoy the experience, he reprimanded several of them for poor judgment and tried to ensure that these sorts of media problems would not be repeated. He also publicly defended Gus Grissom, whom he disliked and privately chastised, over the famous "deli sandwich" episode, in which the astronaut, along with crewmate John Young, ate a smuggled corned beef on rye on the Gemini III flight in 1965. NASA's administrator, James Webb, passed the licks for such misbehavior to Gilruth, writing after the Grissom incident:

> If this were a military operation and this kind of flagrant disregard of responsibility and of orders were involved, would not at least a reprimand be put in the record? . . . The only way I know to run a tight ship is to run a tight ship, and I think it essential that you and your associates give the fullest *advance* consideration to these matters, rather than to have them come up in a form of public criticism which takes a great deal of time to answer and which make the job of all of us more difficult.

None of this suggests that NASA officials let the astronauts run amok, of course, but they did try to maintain order through means more patriarchal than military.

As often as not, the astronauts' wives also enforced considerable discipline on their husbands. The public persona of these women was always "proud," "thrilled," and "happy," their watchwords for the pride they presumably took in the fact that their husbands had been given the greatest opportunity ever enjoyed by a group of military pilots. Their private well-being, however, was always something less as they struggled with home and husband. The public saw them spending their time caring for their families and one another and supporting the efforts of NASA in reaching for the Moon.

The astronauts' wives formed their own little clique in Houston in the early 1960s. Originally the group included just the wives of the first

Mercury Seven astronauts, led by "Mother" Marge Slayton, wife of Deke, who commanded the women with an authority that even her husband must have envied. They transferred to this new setting the lifestyle they had developed during their military experience, replicating its social networks, hierarchies, and priorities at NASA. Anyone who has ever spent any time on a military base understands the responsibilities and authority of the wives of senior officers: They set the standard for the moral and social well-being of the base, serve as a support for group members in both good and difficult times, and enforce the principles of the organization. That is exactly what the astronauts' wives did.

They also locked arms in situations when threats to the group arose, especially when any of its members threatened to upset the balance of their lives. They could turn, seemingly at a moment's notice, to offer aid and comfort or sanction and censure as directed by the leaders of the group. Their concerted actions took two forms, the first seen when one of their number suffered the loss of a husband. Astronauts engaged in a risky profession, as did the military test pilots from whose ranks the astronaut corps was drawn, and some died in the performance of their duties. Charlie Bassett and Elliot See were killed in an airplane crash in February 1966, and Gus Grissom, Ed White, and Roger Chaffee died in the Apollo 1 ground test fire test in January 1967. In both instances, the wives swung into action to care for the bereaved families, to offer support to the widows, and "to maintain an even keel" (a nautical term used by Alan Shepard in many situations). This was a situation they knew well from their experiences with their husbands' active-duty flying.

Second, in another situation they had encountered in their military experiences, the wives of the astronauts had to deal with their husbands' infidelities. The opportunities for cheating were greater for the astronauts than for most others at the time: they spent a lot of time at Cape Canaveral each week, leaving their families behind in Houston. As celebrities, they found themselves pursued by women at every turn. Some of the men lost sight of their marriage vows. Indeed, some, such as Grissom, were notorious womanizers. He was not alone, though. When Apollo 7 astronaut Donn Eisele left his wife, Harriet, to marry another woman whom he had been seeing for some time, the divorce pulled the covers off a longtime practice. The other wives surrounded Harriet with loving support and ostracized the second wife. It did not take long for Eisele and the scandal tied to him to be exorcised from the community, for he soon left the astronaut corps.

At the same time, much of the response to the infidelity involved women talking about the husbands of others, not their own experiences. This denial may have served the peace of mind of the aggrieved wife, but it suggests that the astronaut corps was more of a Peyton Place than anyone seemed willing to admit at the time. This idea is supported by the fact that a large number of the couples divorced, and it appears that some stayed together as long as they did only out of a sense of responsibility and their determination not to embarrass NASA.

In only a few instances did the astronauts violate sufficiently serious rules to warrant public punishment from NASA. For example, in 1971, Apollo 15 mission commander David Scott took, without authorization, 398 commemorative postage stamp covers to the Moon with the under- standing that a German stamp dealer, Hermann Sieger, would purchase them on return and then resell them as flown items, thereby enhancing their value. As a moneymaking scheme, this was not a supervillain-level crime. There was only one source from which the memorabilia could be obtained, and as soon as the stamps went onto the market, NASA moved to punish those involved. As Jeff Dugdale of *Orbit* magazine commented:

> David Scott, the Commander, was famously dismissed from the Astronaut Corps on the first anniversary of his return from this mission as the Apollo 15 crew had smuggled 400 space covers with them. It was reported in newspapers in July 1972 that a West German stamp dealer had sold 100 of these at £570 each. Each of the three crew members had been expected to gain as much as £2,700 from the sale of covers. However they then declined to accept any money, acknowledging that their actions had been improper. [Crewmates] Jim Irwin also resigned from the Astronaut Corps, and [Al] Worden was also moved out of the select group and made no more flights.

The scandal had a long tail. In 1983, Scott sought to recover 298 of the flown stamp covers that had been confiscated after the mission, arguing that he had only violated agency rules and had done nothing illegal. "We were reprimanded and took our licks. But it was a very raw deal," he recalled.

This was a rare instance in which public shaming and cashiering took place during the Apollo program. NASA could not punish astronauts, heroes in the eyes of the American public, without an airtight reason. Bob Gilruth characterized it as trying to keep issues in perspective: these men put their lives on the line, and so, NASA reasoned, they deserved some leniency when problems arose. After all, they repeatedly rose to the challenge in conducting the Mercury, Gemini, and Apollo missions.

Although the astronauts occasionally bucked NASA's restrictions, because of their backgrounds as military test pilots, they understood very well how to thrive in a large bureaucracy with divergent cultures and myriad priorities, and for the most part they successfully negotiated organizational shoals. They also understood that the piloting skills that were emphasized during the Apollo program eventually would grow less important to NASA than scientific and engineering prowess. Fliers and flier culture dominated the Astronaut Office for a time but would not be advantageous for the agency's future; what NASA really needed were scientists who could undertake field research on the Moon and in other settings in space. Pilots for space vehicles would always be needed, but the astronauts recognized that NASA would need fewer of them in the long term than it would need of scientifically skilled crew members who could further the agency's critical goals. Eventually a few pure scientists would be selected for the corps, and one, Harrison Schmitt of Apollo 17, would even go to the Moon.

The result was something of a love/hate relationship between the astronauts and their employer. Every non-astronaut at NASA recognized the value of the astronauts for public relations, and the astronauts, realizing they needed to remain beloved resources for NASA, policed themselves within certain bounds to keep hell-raising under control. They became, as a result, some of the first professional organization men, somewhat like advertising executives or other highly skilled professionals, embracing the larger culture of the organization in which they labored.

For NASA's part, agency officials basked in the reflected glory of the astronauts. Gilruth, among others, enjoyed being with them and loved their humor and passion for life. He said, "People used to tell me that

I had no control over the astronauts. I'll tell you, those boys were wonderful." All the astronauts epitomized the theme of American manhood, which by this time could be characterized as all-American boyishness, their impulses being their best part. The astronauts put a very human face on the grandest technological endeavor in history.

At the same time, the myth of the virtuous, no-nonsense, able, and professional astronaut was born with the space race in the 1960s. In some respects, it was a natural occurrence. Those who went to the Moon seemed to be Everyman. None was aristocratic in bearing or elitist in sentiment, and few had privileged backgrounds. They came from seemingly everywhere in the nation, excelled in public schools, trained at their state universities, served their country in war and peace, married young, tried to make lives for themselves and their families, and ultimately rose to their places based on merit. Diligence, achievement in school, excellence in math and science, and an unfaltering devotion to duty were the prerequisites. Family men with wives and children and military veterans who attended college on the GI Bill, they seemed to be perfect all-Americans. Like the myth that any American child is able to grow up to become president, the astronauts suggested that anyone could become an astronaut. They represented the best we had to offer, and, most important, they expressed at every opportunity the virtues ensconced in the democratic principles of the republic.

Some astronauts bristled at NASA's efforts to cast them as being pure as the driven snow. Apollo 7's Eisele ran afoul of James Webb and made clear he had no use for the administrator, calling him a "sorehead" and "stuffy as hell." Eisele's posthumously published memoir concluded that Webb "wanted the entire agency to be faceless, amorphous, impersonal—except for him, of course." Still, the image of the astronaut remained relatively stable until the 1977 publication of Walter L. Cunningham's autobiography, *The All-American Boys.* Cunningham, lunar module pilot on Apollo 7 in October 1968, not only captured what it was like to be at NASA during Apollo but also effectively addressed what the astronaut corps was about and revealed how it operated as a surrogate for all Americans in the Cold War era. Cunningham was on the periphery of NASA's astronaut corps, never especially well known or promoted by NASA as one of its quintessential heroes, and his book explores the politics of the Astronaut Office, the competition among astronauts for flights and favor, and the difficulties of interrelations in a group of exceptionally accomplished overachievers who sported overactive egos and sex drives. Cunningham allowed Americans behind the curtain that had masked the reality of the Apollo astronauts.

Many have followed Cunningham in exposing the Apollo astronauts to public scrutiny, some revealing (or repeating) stories of sexual impropriety. For instance, Grissom may have had an illegitimate child. "Rumors surrounded Grissom," wrote journalist James Schefter, "including the unproven rumor that he fathered an out-of-wedlock child born to a secretary at the McDonnell Aircraft Corporation in St. Louis." Yet the public response was muted; most people took the revelations in stride.

Political cartoons published during the Apollo era provide a useful window onto public perceptions of the astronauts. In 1969, the year of the first Moon landing, Apollo was omnipresent in the news and was the topic of cartoons throughout the year. In the period January 1 to July 18, many cartoons mentioned astronauts Gene Cernan, Tom Stafford, John Young, Neil Armstrong, Buzz Aldrin, and Mike Collins by name, and most placed the men in a heroic context, some implying that their accomplishments had contributed to national pride. Social scientist Kathy Keltner's analysis of 1,169 political cartoons mentioning specific Apollo astronauts has found that they were depicted as heroes in 42 percent of these cartoons, while the rest presented them neutrally; only 8 percent portrayed them in any negative fashion. Additionally, the astronaut was presented expressly as an ordinary American in 18 percent of the cartoons. Keltner notes that

> the general trend of all of these cartoons seems to be that
> mentions of astronauts as hero decreased as Apollo got to be
> "old news" after 1969. Not only did mentions of actual names
> decrease over time, but so did astronauts being depicted as heroes.
> Interestingly, as mentions of names and hero status decreased,
> astronauts were increasingly being placed in symbolic metaphor
> contexts as time elapsed, both in political and comic strip
> cartoons. . . . Between 1961–1973, political cartoons continued to
> take pride in the work of astronauts, and realized the importance
> of their missions. . . . Because the technology was so new, and

many thought it hard to believe man could actually go to the moon, these cartoons may have placed the Apollo program into cultural contexts, creating symbols shared by society in order to understand the reality, significance and importance of the missions—at least for those who read comic strips. Those who read political cartoons are more likely to be politically versed and more educated, leaving the general public to unpack the cultural metaphors of comic strips that entertain and enlighten through shared symbolism.

Examples of the heroizing dynamic abound. For example, the life of Alan B. Shepard Jr., the first American in space in May 1961 and commander of the Apollo 14 mission in 1971, was presented by the press and NASA alike as the story of an Everyman. Born in 1923 in East Derry, New Hampshire, he was educated in public schools and at the US Naval Academy in Annapolis between 1941 and 1945, becoming a career officer and serving first on a destroyer and later as a naval aviator. His later exploits as an astronaut were among the most significant scientific endeavors of the recent past. When he became the first American to ride a rocket into space, he became both a celebrity in the public's eyes and, due in part to NASA's deft public relations, a frontiersman in the mold of Lewis and Clark. His actions as one of the agency's "point men," as it called the seven Mercury astronauts, helped to unify the nation behind the exploration of space. His suborbital Mercury mission established that the United States could send an individual to space and then recover him, an enormously significant event for a country hoping to recover its honor in the wake of shocking Soviet achievements, Sputnik in 1957 and Yuri Gagarin's space flight earlier in 1961. The flight made of Shepard a national hero, but his stoic persona and public countenance also served to solidify his stature among Americans as a role model.

Shepard's other great space flight took place a decade later, from January 31 to February 9, 1971 (a medical disorder had kept him off flight status for several years), when he commanded Apollo 14. His mission occurred at a tense time, just a few months after the near-tragic Apollo 13 mission, and its complete success boosted the national spirit and restored faith in the Apollo program. The achievements of Apollo 14 were many: the first use of the mobile equipment transporter; the placement of the largest payload ever in lunar orbit; the longest stay on the lunar surface (33 hours); the longest lunar surface EVA until that point (nine hours and 17 minutes); the first use of shortened lunar orbit rendezvous techniques; the first use of color television on the lunar surface; the first extensive orbital scientific experimentation period conducted in lunar orbit; and even the first lunar golf game, in which Shepard, an avid golfer, hit a hole in one. Shepard's essential humanity came through in this effort, as he accepted his role as both an American icon and an ordinary citizen who was simply answering the call of his nation.

Frank Borman was born and raised near Gary, Indiana, in 1928, and his story is much like Shepard's. He attended public schools, graduated from the US Military Academy at West Point in 1950, and entered the Air Force, where he became a fighter pilot. From 1951 to 1956, he was assigned to various fighter squadrons. After completing a MS in aeronautical engineering, he became an instructor of thermodynamics and fluid mechanics at West Point in 1957. On September 17, 1962, Borman became an astronaut with NASA. He would command the Gemini VII mission in December 1965 and take part in the longest space flight up to that point (330 hours and 35 minutes) as well as the first rendezvous of two maneuverable spacecraft. Borman's most significant space mission was as commander of the Apollo 8 mission, which flew around the Moon over the Christmas holiday in December 1968. An important accomplishment, this flight helped to draw the country together at a time when American society was in crisis over Vietnam, race relations, and social unrest.

Most prominent among the Everyman heroes was Neil Armstrong, born on August 5, 1930, on his grandparents' farm near Wapakoneta, Ohio. He developed an interest in flying at age two, when his father took him to the National Air Races in Cleveland. His interest increased after he had his first airplane ride in a Ford Tri-Motor, a "Tin Goose," in Warren, Ohio, at age six. From that time on, he had an intense fascination with aviation. At 15, Armstrong began taking flying lessons at an airport north of Wapakoneta, working various jobs in town and at the airport to earn money for lessons in an Aeronca Champion. By 16, he had his student pilot's license, even before he had passed his driver's test and graduated from Wapakoneta's Blume High School in 1947.

Immediately after high school, Armstrong received a scholarship from the US Navy. He enrolled at Purdue University and began to study aeronautical engineering. In 1949, the Navy called him to active duty, and he became an aviator. In 1950, he was sent to Korea, where he flew 78 combat missions from the aircraft carrier USS *Essex*. After mustering out in 1952, Armstrong joined the National Advisory Committee for Aeronautics (NACA). His first assignment was at NACA's Lewis Research Center near Cleveland. For the next 17 years, he worked as an engineer, test pilot, astronaut, and administrator for NACA and its successor agency, NASA.

In the mid-1950s, Armstrong transferred to NASA's Flight Research Center in Edwards, California, where he became a research pilot on many pioneering high-speed aircraft—including the well-known X-15, which was capable of achieving a speed of 4,000 miles per hour. He flew more than 200 models of aircraft, including jets, rockets, helicopters, and gliders, pursued graduate studies, and received an MS in aerospace engineering from the University of Southern California. Armstrong transferred to astronaut status in 1962, one of nine NASA astronauts in the second class to be chosen.

On March 16, 1966, Armstrong flew his first space mission as command pilot of Gemini VIII with David Scott. Armstrong piloted the spacecraft to a successful docking with an Agena target spacecraft already in orbit. Although the docking went smoothly and the two craft orbited together, they began to pitch and roll wildly. Armstrong undocked the Gemini and used retro rockets to regain control of his craft, but the astronauts had to make an emergency landing in the Pacific Ocean.

As spacecraft commander for Apollo 11, the first piloted lunar landing mission, Armstrong gained the distinction of being the first person to land on the Moon and the first to step on its surface on July 20, 1969. The Apollo 11 astronauts were honored with a tickertape parade in New York City soon after returning to Earth. Armstrong received the Medal of Freedom, the highest award offered to a civilian, as well as the NASA Distinguished Service Medal, the NASA Exceptional Service Medal, the Congressional Space Medal of Honor, and 17 medals from other countries.

The experience of these astronauts is emblematic of all those who went to the Moon during Project Apollo. Their backgrounds and career paths were similar and representative of the white male experience in America. Was this an intentional choice by NASA, or was it serendipity?

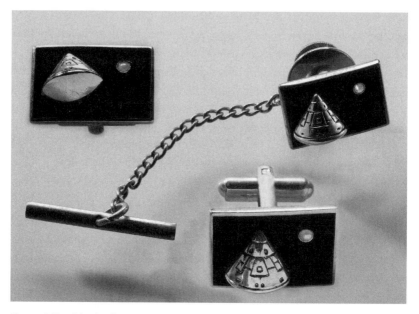

Memorabilia of the Apollo program included high-end tie tacks and cufflinks such as the commemorative set shown here, which was given to Apollo 11 astronaut Michael Collins. (Photograph by Eric Long, NASM image no. 19-2009_4782)

Probably it was a bit of both. Overall, the Everyman template served as an important consensus-building element in the national space program. Most Americans—whether liberal or conservative—could identify with middle-class astronauts who had been educated at state universities or military service academies and who seemed to embody the qualities viewed as important to national well-being.

Nothing points up this consensus more effectively than a political cartoon published at the time of the Apollo 17 lunar mission in December 1972. It shows two African Americans watching Gene Cernan and Harrison Schmitt on the Moon; the caption reads, "Maybe we'll go next time." But Apollo 17 was the last of the Apollo missions to the Moon, and no one has returned since. By highlighting the exclusion of the African American community from this epochal event, the cartoonist emphasized the Anglo-Americanness of the entire episode. It was, without a doubt, both a mainstream American experience and one that marginalized racial and ethnic minorities even as it solidified traditional national virtues and perspectives.

This purse in the shape of the Apollo spacecraft was one of several items created to commemorate the Moon program. Such items were popular in the later 1960s and early 1970s and fit well with a couture that emphasized human-made fabrics and space-age ideas of fashion. (Photograph by Eric Long, NASM image no. 2005_36250.09)

The whiteness of NASA and the marginalization of African Americans within the space agency are also exemplified by the story of Robert H. Lawrence Jr., the first African American selected to be an astronaut. Born in Chicago on October 2, 1935, he excelled in school and received a PhD in physical chemistry at Ohio State University in 1965. Already a pilot in the US Air Force, he entered the Air Force's Manned Orbiting Laboratory (MOL) program in June 1967. Before he could fly in space, Lawrence died in an F-104 crash on December 8, 1967. Not until 30 years later would NASA, which had never officially recognized Lawrence as an astronaut, place his name on the Astronauts Memorial Foundation's Space Mirror at the Kennedy Space Center in Florida, the 17th such enshrinement.

Wherever astronauts have gone, from the beginning of the spaceflight program to the present day, their uniforms have characterized and embodied them. In both fact and fiction, an astronaut's spacesuit has been

Space-themed Thermos and lunchbox sets became popular during the space race of the 1960s. This one features multistage rockets, wheeled space stations, and a human base—images that closely resemble pop conceptions of spaceflight from the 1950s. Over time, many lunchbox designs became more closely related to the Apollo Moon-landing program, and their popularity reached its high point with the landings of the late 1960s. (Photograph by Eric Long, NASM image no. 15-2005-36194.46)

a core representation of the person, essentially a medieval knight's armor worn while the individual conducts his noble mission. As the astronauts walked on the Moon, with their visors down and their identical suits making them anonymous, they seemed to be mysterious and attractive archetypes, conjuring up images of power and masculinity far beyond what viewers actually saw. Without NASA's intending it, the spacesuit, more than any other Apollo artifact, became synonymous with a set of heroic values dominating what Americans wished to believe about themselves and their nation. It seemed to manifest the ideals that had supported Americans' going into space in the first place, symbolizing and serving as a metaphor for the utopian desire to colonize the solar system and make a perfect society in a new and pristine place beyond the corrupt Earth.

An African American technician at a wind tunnel test of the Saturn I at the Langley Research Center, Hampton, Virginia, March 2, 1962. There were few minority workers at NASA at the time, but the space agency staged this image with an unnamed technician to demonstrate its commitment to equal opportunity. (NASA image no. L-1963-01637)

At the same time, the Apollo astronauts, in their visored white space-suits, became screens onto which the whole of America could project its hopes, wishes, fears, and horrors. Each astronaut felt this projection keenly, and they have lived out the remainder of their lives in the glare of American fame and a sense of expectations never fully satisfied. Unable to always reflect the qualities of strength, authority, and rationality projected upon them, they have sometimes shown a fragility over the decades since Apollo that has perplexed or troubled onlookers. Writer Marina Benjamin captures this sense best when she tells of encountering three Apollo astronauts at a collectors' show: they were "just like movie stars; they burned brightly in the glare of publicity when they were offered good parts to play and then, when the roles dried up, so did they." Their spacesuits, however, have endured. Suggesting the triumph of technology over living organisms, they continue to take center stage in museums and science centers around the nation.

In their day, the Apollo astronauts also epitomized a hardworking, fun-loving, virile idea of the American male. US public comfort with the white male establishment is palpable whenever stories of Apollo are recounted: in such tellings, the astronauts were quintessential company men who happened to work for NASA during Apollo, and the engineering "geeks" of Mission Control, with their short-sleeved white shirts, narrow black ties, slide rules hung on their belts like sidearms, and pocket protectors complete with compass, ruler, pens, and mechanical pencils, personified a conservative America that many still look back on with fondness and nostalgia.

Norman Mailer, as much an embodiment of the 1960s counterculture as anyone, noted this aspect of Apollo while covering the Moon landings in 1969, expressing fascination and not a little perplexity with the milieu he encountered at the Manned Spacecraft Center in Houston. On the one hand, he railed against overwhelmingly white male NASA, steeped in middle-class values, reverence for the American flag, and mainstream culture. On the other, Mailer grudgingly admitted that NASA's approach to task accomplishment—which he viewed as the embodiment of the Protestant work ethic—and its technological and scientific capability got results with Apollo. For all his skepticism and esotericism, and even though he hated NASA's closed and austere society, which he believed distrusted outsiders and held them at arm's length with a bland and faceless courtesy, Mailer's writing captures much of interest concerning rocket technology and the people who produced it during Project Apollo.

Calling himself Aquarius, Mailer, in his 1970 book *Of a Fire on the Moon*, describes in poetic, exalted language the launch of the mighty

Moon rocket that carried the Apollo 11 crew beyond Earth. Since light travels faster than sound, he sees the engines fire before he hears them, or feels them, and he comments that the Saturn V liftoff seems to be "more of a miracle than a mechanical phenomenon, as if all of the huge Saturn itself had begun silently to levitate." He notes that the five F-1 engines blaze in "brilliant yellow bloomings of flame," turning "white as a torch and as long as the rocket itself." At this point, the sound waves reach him like the "thunderous murmur of Niagaras of flame roaring conceivably louder than the loudest thunders he had ever heard and the earth began to shake and would not stop." Mailer experiences what he considers to be the full transcendental nature of this launch, commenting that the rocket looks "like a ball of fire, like a new sun mounting the sky, a flame elevating itself," and that finally humanity "now had something with which to speak to God."

Film as well as literature helped to mythologize the Apollo missions. In 2004, NASA studied representations of astronauts in movies and found that the majority have celebrated their skill and masculinity, especially films focusing on aspects of the Apollo story:

> As a group, the public entertainments we tend to buy into are either nostalgic visions of the "space race" period ("The Right Stuff," "Apollo 13," "From the Earth to the Moon") or fantasies reflecting the romantic imagination of the Flash Gordon/Buck Rogers era ("Star Wars" rather than "Star Trek"). These are the visions people support in the most meaningful way possible: with their time and dollars. . . . Boomers have a great nostalgic affection for NASA, but their own priorities have shifted from a future focus to maintaining what they have. They see money spent on space exploration as threatening their future entitlements.

At a sublime level, the Apollo astronauts may serve as a symbol of a decreased interest in the future expressed by Americans in the first decades of the 21st century. The country's shifting cultural center of gravity—toward maintenance of the status quo and away from looking to the future—has clouded the shared national vision that energized earlier space efforts. Accordingly, we might glory in the success of Apollo, but we do not embrace a return to it and its risks.

Yet even from the perspective of the 21st century, the astronauts of Apollo, indeed all astronauts anytime, retain a special place in the American experience. They have come to exemplify excellence, perseverance, victory in the face of adversity, and what many Americans consider to be traditional values that made the United States a nation favored among all others. This sense of "chosenness" might have religious overtones, but it also expresses the secular ideal of America as a land of opportunity where all may achieve their proper rewards through diligence and hard work. World Wars I and II had led Americans to believe they were fighting for the survival of all that was good against forces of evil. The context of the Cold War made it easy for Americans to perceive the astronauts as proxy soldiers continuing that struggle on the nation's behalf, defending it from the ravages of a stalwart rival and serving as surrogates for their society. Historian Richard T. Hughes has observed that Americans tend to see the country as inerrantly "standing clearly and unambiguously on the side of the right," although, of course, "the world does not, in fact, divide as neatly between good and evil as the myth of America as the Innocent Nation might suggest."

Arising during the Mercury era and sustained by the astronauts during the heroic age of Apollo, the mythology of space exploration has remained amazingly fixed since it took root. Astronauts' membership in an elite "club," supported by ritual and secrecy, did much to establish their group identity and maintain a boundary between them and the rest of society, as did NASA's and the media's celebration of their skill, masculinity, ordinary backgrounds, patriotism, hard work, and extraordinary space accomplishments. Although the initial ranks of white male astronauts eventually would open to more Americans on the basis of merit, they have remained an elite group. Without question, they have served well the larger purposes of American greatness, both in the 1960s and afterward.

The astronauts, individually and collectively, have remained in the heroic mold. Together they are one of the truly great examples of myth-making in modern American history. All astronauts are viewed as virtuous and heroic, cool under pressure, and technologically dexterous. They are brilliant and attractive, impressive in every setting. They are able, bold, learned, and brave. They exude Americanness and patriotism of the highest caliber from every pore. Could any of them ever go wrong?

Perhaps not, since they are sought after like only a few other celebrities. Two personal anecdotes demonstrate this point.

In early June 1999, I attended a formal dinner at the annual National Space Forum sponsored by the American Astronautical Society. Seated at the table were nine other people, one of whom was Frederick Hauck, an astronaut on three Space Shuttle missions and commander of STS-26, the first mission to fly after the *Challenger* accident, by then retired from NASA and leading an aerospace company. Hauck's name, like those of many astronauts of the shuttle era, is unknown to most people, and no one at the table except for me seemed aware of his earlier career. As we talked, however, he casually mentioned one of his shuttle flights, whereupon those at the table began fawning over him and asking for his autograph. Hauck and I discussed this afterward and agreed that it was the height of mythmaking that one could be transformed, in the twinkling of an eye, into a hero merely by being identified as an astronaut.

On July 20, 1999, NASA celebrated the 30th anniversary of the first lunar landing, and the crew of Apollo 11—Neil Armstrong, Buzz Aldrin, and Michael Collins—came to Washington, DC, for the festivities. Among the other rather tiresome autograph seekers was an important congressman demanding of the NASA administrator that he be allowed to bring his twin daughters to meet the crew, have their pictures taken with them, and get their autographs. NASA officials put him off repeatedly, in no small measure because of the crew's desire not to be hounded, but he showed up unannounced at an event anyway, without his daughters (whom he said were sick that day), and demanded to be photographed with the astronauts and get their autographs. The crew, confronted with the aggressive congressman, agreed to the photograph but not autographs. Armstrong, Aldrin, and Collins are genuine heroes, but it is curious that a respected national political leader would blatantly use his power for such a thing. The only explanation is the overpowering perception that astronauts, especially the revered Apollo 11 crew, are iconic.

That aura remains, even as the astronauts grow older and their presence becomes less impressive. Meeting Apollo astronauts still alive in the 21st century inevitably points up the difference between our perceptions of them in their prime and their current elderly lives. Indeed, most of those who remember Apollo astronauts have frozen them in time as youthful heroes despite the many years since the Moon landings.

Some have passed on. Interviewed in 2006, Charlie Duke, a member of the Apollo 16 crew, touchingly recalled that only a few moonwalkers were left. Not too far in the future, he sadly noted, none would remain. The survivors are now in their 80s, and while some remain in good shape, the rigors of long lives are evident. The brash, swaggering, virile, accomplished, death-defying pilots of the Apollo program are now stooped and time-ravaged. As journalist Andrew Smith writes, they are no longer "the most perfect imaginable expression, embodiment, symbol, of the 20th century's central contradiction: namely, that the more we put our faith in reason and its declared representatives, that the more irrational our world became."

Others, especially space program critics on the left, have theorized that the Apollo astronauts were representative of an emerging "ornamental culture" at NASA in which the astronauts were little more than props for a larger American publicity campaign. "The astronaut served as an emblem in many matters preoccupying cold-war America: beating the Russians, demonstrating national mastery, wedding technology to progress, proving the point of man over machine. But paramount among his symbolic roles, he was to be a masculine avatar for a strange and distinctly new realm on earth," writes cultural critic Susan Faludi, who contends that the astronaut was "a first-draft response to disturbing questions about manhood in an ornamental age." Rather than being valued for their capabilities in pushing back the final frontier, Faludi comments, the astronauts were charged with the opening of a new entertainment frontier. In that sense, she draws direct linkages between the astronauts and earlier entertainments such as Wild Bill Hickok's Wild West Show. "But the astronauts heralded a time," she emphasizes, "when the sideshow would as never before supplant the main event." All of this was entirely understandable, Faludi adds: "NASA needed the pleasing faces, the frenzy of celebrity, to seduce the government, the media, and the public into accepting the huge expense of the aerospace program."

But were the astronauts simply ornamental? Clearly they were celebrities, but that status seems to have been predicated on their exciting and important work. Other interesting questions arise: Were the astronauts famous for simply being famous? Were they famous for "real" feats or only perceived ones? Did the public really understand (or care) about their achievements? Or were they famous because somebody told the public they were famous? The astronauts, more fairly, could be likened to sports and entertainment idols. Like them, they had to accomplish great feats to reach heroic status. Yet unlike sports or entertainment figures, their heroism did not truly fade with time, which raises yet another question: How effective was NASA in scripting public perceptions of the astronauts? If their lasting significance is any measure, NASA was very successful indeed.

How might we interpret the astronauts of the Apollo program today? They were truly heroes in a vacuum, but ones who also held a powerful sway over modern American culture. While critics on both the right and the left have questioned whether NASA took too many risks with the lives of the spacefarers—and challenged the program after some died performing their duties—the astronauts themselves carried on, completed their work, and walked in triumph on the Moon.

5

Ex Luna, Scientia

"Okay. Now let's go down and get that unusual one. Look at the little crater here, and the one that's facing us. There is this little white corner to the thing. What do you think the best way to sample it would be?"

Apollo 15 commander David Scott had taken to the geology training he had undergone before flying to the Moon in 1971. He wanted to bring back scientifically useful lunar samples that would help to show how the Moon was formed and and how it had evolved since the beginning of the solar system. Now he and Jim Irwin were on the rim of Spur Crater, about 50 meters above the mare surface, on the slope of Hadley Delta. Irwin commented on a glint from the surface. "Oh, boy!" replied Scott. "Guess what we found? Guess what we just found?" he said to the geologists who were following the moonwalk from a room attached to Mission Control in Houston. The astronauts had discovered what they came for: a rock that might hold the answer to the question of the Moon's formation.

Later, during a trans-Earth press conference, scientists referred to this lunar sample as the Genesis Rock, a name that has stuck. The anorthosite sample, number 15415, proved to be more than 4 billion years old, formed in the early stages of the history of the solar system and therefore providing a window into the origins of the Moon, Earth, and our solar system. As Scott described the treasure to the media:

I think the one you're referring to was what we felt was almost entirely plagioclase or perhaps anorthosite. And it was a small fragment sitting on top of a dark brown larger fragment, almost like on a pedestal. And Jim and I were quite impressed with the fact that it was there, apparently waiting for us. And we hoped to find more of it, and, I'm sure, had we more time at that site, that we would have been able to find more. But I think this one rock, if it is, in fact, the beginning of the Moon, will tell us an awful lot.

And we'll leave it up to the experts to analyze it when we get back, to determine its origin.

The astronauts had correctly recognized the importance of their discovery. Scientists had worked hard to ensure that the flight crews had the knowledge necessary to undertake useful work on the lunar surface. To a surprising degree, they succeeded. Between 1964 and the times of the various missions, the Apollo crews undertook classroom study and fieldwork in a variety of settings to prepare for their time on the Moon, accumulating sufficient formal education to earn the equivalent of a master's degree in geology.

Indicative of the approach taken by some of the astronauts was Scott's work on Apollo 15. He enthusiastically trained for the mission, and once on the Moon he concentrated on scientific efforts. "Most of my thoughts on the Moon were of the geology involved," he later recalled. "Our mission was especially heavy in science, trying to understand the geology of the local site . . . why things occurred as they did."

The story of the Genesis Rock was a high point in the science conducted on the Moon. "The direct scientific result of the Apollo Program, viewed collectively, can be summarized as fundamental new knowledge of the Moon, the Sun, and the Earth, and of the behavior of living and inanimate systems in the microgravity environment provided by orbiting spacecraft and space stations," according to longtime NASA scientist Paul D. Lowman. In a triumphalist statement on Apollo's scientific harvest, lunar geologist Don E. Wilhelms said that it was "a once-in-a-lifetime opportunity" to learn about our place in the solar system, and not to have taken full advantage of it "would have been unconscionable."

Even so, the core reasons for undertaking Apollo had little to do with furthering scientific understanding. Its impetus rested, as noted earlier, almost solely on Cold War rivalries and the US desire to demonstrate technological superiority to all the peoples of the world. Nonetheless, a great amount of scientific knowledge emerged from the missions to the Moon. Ironically, this harvest of information, while fully incorporated into the scientific literature and proving to be a transforming force in our understanding of the cosmos, may not have made much sustained impact on general perceptions of the Moon itself, which remains a metaphysical entity to many people (as evidenced by the rise in recent decades of full-moon spiritual gatherings, planting by moonlight, and the like). A question to be considered is whether the demystification of the Moon by the Apollo program was followed by a remystification, or whether it never truly became demystified at all.

The science of Apollo is of considerable significance in each of the four main threads of our recollections of the program. It looms particularly large in the dominant narrative of American triumph, and has been employed as a justification for the lunar landing effort beyond the geopolitical objectives that motivated the 1961 decision. It serves the larger purpose of offering to the world something beyond a demonstration of American technological virtuosity. Emphasis on the scientific fruits of the missions developed gradually as scientists worldwide were afforded the opportunity to study Apollo's geological samples and the data gathered on the missions and as scientific textbooks were rewritten to accord with the new findings.

In the other main threads of the Apollo story, those who criticize the Moon landings from both the left and the right answer the question of whether its toil, trouble, and especially cost were worthwhile in the negative. Advocates of these perspectives argue that the funding for the program could have been more effectively used elsewhere and that the scientific results did not justify the investment. As journalist Andrew Smith remarked in a book tracking the Apollo astronauts since the program ended, "Was Apollo worth all the effort and expense? If it had been about the Moon, the answer would be no." Scientists not involved in space studies especially bemoaned the expenditures made on Apollo, and even some space scientists thought their field would have reaped more benefits had the funds been directed elsewhere, especially toward robotic exploration of the Moon and the planets. NASA's senior scientist Homer Newell recalled that during a 1962 review of the program, "Many [scientists] expressed disapproval of the manned program, along with the wish that the monies going to Apollo might be diverted to space science. Some expressed concern that not only was Apollo going to proceed but that NASA would even seek to justify the program on the basis of science, and this the scientists strongly objected to." As one wit suggested, "We were manned if we did, and manned if we didn't."

For those on both the left and the right, the cost versus scientific gain argument offered a powerful critique of Apollo. The contention that Apollo was enormously expensive, and that the funding could be better used to address other needs, proved persuasive to many. Critics of lunar science used the same argument, suggesting that science accomplished through other means would be more cost-effective.

In addition, some scientists on both the political left and right opposed Apollo because it lassoed science into the service of the state. Some on the left believed this was an inappropriate use of science, especially the manner in which it drew space science into the military-industrial complex and corrupted the pure quest for knowledge. Merle Tuve, of the Carnegie Institution of Washington's Department of Terrestrial Magnetism, was a longtime opponent of the cooptation of science and scientists for government, especially military, purposes. Apollo science was simply one more instance, he believed, of government big-footing scientific investigation, which he'd seen repeatedly in the era since World War II. Willingness to accept government funding for applied research, Tuve said, could "reduce the future opportunity for devoted studies of the intricate mysteries of nature." His purist sensibilities equated Apollo with military research, and military research was something to be avoided at all costs. Likewise, Columbia University's Nobel Prize–winning physicist Polykarp Cush, a social liberal who embraced the reforms of the 1960s, believed that the money lavished on Apollo science was an inappropriate expenditure. He did not oppose the Moon landings altogether, but he viewed them mostly as an overreach at a time when the government should have been taking action to deal with inequities in society.

Meanwhile some scientists on the political right also voiced concern about Apollo, largely because they saw it as a huge government program that replaced individual accountability for the conduct of investigations with state support, authority, and management. For them, this represented a massive intrusion into what they conceived of as the pure pursuit of knowledge of the natural universe by a government-constructed and -controlled master plan. One editorial, picking up on such critiques, remarked on the 30th anniversary of the first Moon landing in 1999 that Apollo had proved a fleeting and hollow triumph:

It is far cheaper and much easier, of course, to fling intelligent robots through the solar system than to send astronauts back to the Moon. . . . Our footprints are on the moon, along with the late Alan Shepard's golf ball and mementos left behind by other astronauts, but our home is here, on the great round Earth, and it is here that we are likely destined by biology to stay. No astronaut who went to the moon ever expressed a desire to stay there, and no robot ever radioed for directions back to Earth.

The emphasis on cost, results, and misplaced effort gained expression through such criticisms, while those who were actually connected to the scientific aspects of Apollo did not effectively make their own case to the public. Most of the time, they celebrated what had been gained without reference to what might have been or to the costs of the effort.

Finally, for those who denied the landings altogether, lunar science from Apollo offered a special challenge. They had to find ways to counter scientific evidence offered by lunar rocks brought back by the astronauts, experiments left on the Moon, and analyses that demonstrated the striking divergence of the Moon from earthly counterparts. For example, in the fall of 2002, Comedy Central's *The Daily Show* produced a segment on denials of the Moon landings, focusing on a recent altercation between conspiracy theorist Bart Sibrel and Apollo 11 astronaut Buzz Aldrin (see page 180). When Sibrel accosted him at a public event, calling him a coward and liar, Aldrin cold-cocked him with a right hook. *The Daily Show* took the opportunity to satirize the episode and interviewed me as one of several spaceflight experts. One tongue-in-cheek question got to the heart of the debate over lunar science and the Moon landings: "What percentage of the so-called Moon rocks are made out of papier-mâché versus the percentage made out plaster of Paris?"

During the Apollo program, astronauts collected a total of 382 kilograms of soil and rock samples from the various landing sites. A scientific consensus quickly emerged that nothing like these rocks existed on Earth; they were easily distinguishable from meteorites coming from elsewhere. Most important, their geochemical characteristics failed to show any indication of water molecules, but they did suggest an impact on an airless body. They were also far older than rocks studied on Earth, some by as much as 700 million years. Finally, they shared the geochemical characteristics of Soviet lunar samples returned by robotic probes. As David McKay, chief scientist for planetary science and exploration at NASA's Johnson Space Center (JSC), explained, "Moon rocks are absolutely unique. . . . Apollo moon rocks are peppered with tiny craters from meteoroid impacts," a feature that rocks on Earth lack.

Their unique composition forced landing deniers to concentrate on trying to find reasons why these samples from the Moon are not really samples from the Moon. Their arguments were both ingenious and ridiculous. A 1994 *Wired* story quoted the denier Bill Kaysing as saying, "The moon rocks were made in a NASA geology lab, right here on earth." He offered no evidence to support this accusation but continued, "Not very many people on the Apollo project knew about the hoax, as they were only informed on a need-to-know basis. Cash bonuses, promotions, or veiled threats could have ensured the silence of those who were in on the whole scheme." Kaysing also argued that because the lunar samples contained no precious metals, they could not be from the Moon—not a statement that supports his theory, since gold, silver, and other precious metals are the result of complex chemical processes that require water on Earth. Kaysing repeated a story he claimed to have heard about a geologist in Washington, who had laughed when someone confronted him with the Moon-landing denial theory and the role of lunar samples in it. For Kaysing, the scientist's reaction proved his point that the rocks were fake; for everyone else, it signaled that the geologist simply was laughing—and who wouldn't?—at the ridiculous conspiracy theory and those holding it. All other efforts to use the scientific results from Apollo to deny the landings themselves have been equally ineffective and absurd. Nonetheless, diehard deniers continue to try.

Even before Apollo, space scientists had a history of riding piggyback on military test programs extending back to the 1940s. For example, after World War II the US Army began using captured German V-2s to learn more about rocket technology. Scientists persuaded Army leaders to allow them to place experiments on the rockets to study the upper atmosphere, and thus, in 1946, the War Department established the Upper Atmosphere Research Panel. Although its name and scope of responsibilities changed periodically during the next several years, it continued to coordinate the placement of scientific payloads on rockets being tested by the military and studied solar and stellar ultraviolet radiation, the aurora, and the composition of the upper atmosphere. As a result, it served as a sort of godfather to the infant field of space science. Scientific data was not the primary purpose of these flights—Army Ordnance mostly wanted to study rocketry to aid development of an advanced generation of weaponry—but nonetheless important findings were gleaned.

Apollo presented scientists with a similar situation: the demonstration of US technical prowess was the primary objective, not scientific study, but room could be made for such work. Some senior scientists preferred not to be coopted by the promise of government funding in exchange for what they assayed as insincere support, and the piggybacking philosophy, in which spending on space science programs presumably rose along with technology demonstration, did not impress them. They had to be convinced.

To make the case, quintessential science entrepreneur Lloyd V. Berkner urged his colleagues in 1961 to recognize that the Apollo program would be completed with or without them and that they might as well take advantage of the substantial funds being made available. In the spring of 1961, Berkner oversaw the preparation of a Space Studies Board report, "Man's Role in the National Space Program," sponsored by the National Academy of Sciences. The report was followed by an SSB press release that declared, "From a scientific standpoint there seems little room for dissent that man's participation in the exploration of the Moon and planets will be essential."

But dissenting voices remained. Lee A. DuBridge, a renowned physicist and presidential science adviser, disparaged Apollo as little more than a prestige project designed to scare the Soviets and impress others. DuBridge possessed a conservative, almost libertarian belief that science would be tainted by government contracts and influence. NASA Administrator James E. Webb tried to convince DuBridge that Project Apollo would yield important scientific data and that scientists should support it, but he never did. Neither did James Van Allen, the dean of astrophysics and a respected voice on behalf of science. Restating the question at the core of his opposition in 2004, he said, "My position is that it is high time for a calm debate on more fundamental questions. Does human spaceflight continue to serve a compelling cultural purpose and/or our national interest? . . . Risk is high, cost is enormous, science is insignificant. Does anyone have a good rationale for sending humans into space?"

Regardless, NASA's space scientists worked to establish a substantial presence for experiments as a part of Apollo, rather than in opposition to it, and they largely succeeded. In 1962, the National Academy of Sciences released a study on scientific justifications for Project Apollo in which two revered space scientists, Eugene Shoemaker and Gerard Kuiper, offered a cogent statement of support: "The lack of an adequate scientific endeavor could invite serious criticism of the program, while the impact of a successful scientific mission by means of a lunar landing will enormously enhance the importance of the Apollo program in the eyes of the world."

They found willing accomplices in favor of using science as a rationale for Apollo among the political elite both in the Kennedy administration and at NASA. *Science* served well as a term that could remove Apollo from the realm of bald-faced geopolitics and wrap it in what most considered a noble pursuit: expanding knowledge for all. The seeds of this approach to spaceflight had been planted many years earlier. As G. Edward Pendray, cofounder of the American Rocket Society, observed in 1945: "The true purpose and fulfillment of life is to know and understand; to see a fuller concept of the world and its place in the universe, and our own position in the cosmic scheme." Space exploration could offer "new knowledge of the realms into which mankind has so far been unable to venture, and thus [it] will stretch our mental horizons and enrich the fields of physics, meteorology, rational and many another science."

NASA Associate Administrator Robert C. Seamans Jr. made the same case in relation to Apollo in late 1961 by emphasizing "the quest for scientific knowledge" as the ultimate reason for undertaking the Moon landings. Webb, his boss, was even more strident, telling President Kennedy that while landing on the Moon was all well and good—it would help demonstrate American preeminence in space—the United States could and should do more. "To be preeminent in space," he said, "we must conduct scientific investigations on a broad front." Likewise, Ernst Stuhlinger, director of the Advanced Concepts Office at NASA's Marshall

Space Flight Center, remarked that "the most profound significance of Project Apollo is its catalytic effect on the material support of large efforts with purely scientific goals."

Some in NASA, however, were skeptical of this argument. One scientist quoted in Donald A. Beattie's retrospective look at Apollo, *Taking Science to the Moon*, succinctly expressed the agency's dominant early thinking on the Moon landings: "We'll land, take a few photographs, pick up a few rocks, and take off as soon as possible." Max Faget, the innovative engineer credited with originating the Mercury spacecraft and playing a leading role in designing the Gemini and Apollo capsules, was more negative than even this. In 1963, he told Joseph Karth, a US representative from Minnesota, that space science should have no role in Apollo, prompting Webb to tell Faget's boss, Robert R. Gilruth of the Manned Spacecraft Center in Houston, to rein in his errant engineer. Faget had failed to understand, Webb believed; he had "missed one of the very important elements necessary to the program. . . . That all he is now doing is based upon scientific work of the past." While NASA engineers might have been more circumspect after Webb's eruption, many remained convinced that the scientific effort was largely superfluous to the Moon landings. As longtime NASA engineer Ivan Bekey remarked, Apollo was in sum an "engineering tour-de-force carrying a large scientific program on its broad coattails."

With many engineers working on Apollo reticent to embrace science as the objective of the program, and many scientists brought to the table hesitantly if at all, it might seem remarkable that any science of significance resulted from the Moon landings. It did, but only because of the dedicated effort of people in both camps who sought common ground. It was always a rocky road to a rocky Moon.

Engineers contended with scientists over Apollo's priorities and prerogatives, with the former focused on the design and flight of the spacecraft and the latter on developing and using scientific instruments. The sides, with their differing goals and cultures, always had an uneasy working relationship as they jostled to ensure that their priorities found appropriate expression. As ideal types, engineers usually worked in teams to build hardware that could carry out the functions necessary to fly the spacecraft to its target and to perform its task once it arrived. Their primary goal was to build vehicles that would function reliably and safely within the fiscal resources allocated to the project. Again as ideal types, space scientists engaged in pure research and were more concerned with designing experiments that would expand scientific knowledge of the Moon. They also tended to be individualistic, unaccustomed to regimentation, and unwilling to concede the direction of projects to outside entities. The competition centered around key mission goals—"Where, for example, would astronauts be allowed to land and for how long? What combination of whose instruments would fly on a given mission? Who would determine how the precious hours on the lunar surface would be divided among doing 'housekeeping' tasks, setting up experiments, and actually exploring the Moon?"—and was expressed in a variety of ways. Scientists disliked having to configure payloads so that they could meet time, money, or launch vehicle constraints. Engineers, likewise, resented changes to scientific packages added after project definition because these threw their hardware efforts out of kilter. Both sides had valid complaints and had to maintain an uneasy cooperation to accomplish all of their objectives.

The scientific and engineering communities within NASA, additionally, were not monolithic, and differences within them thrived. Added to these were representatives from industry, universities, and research facilities, and competition on all levels to further specific scientific and technical areas resulted. NASA leadership generally viewed this pluralism as a positive force within Apollo, for it ensured that all sides aired their views and honed their positions to a fine edge. Competition, most people concluded after the fact, if not in the heat of battle, made for a more precise and viable Moon-landing effort. There were winners and losers in this strife, however, and sometimes ill will was harbored for years afterward. As one Caltech scientist remarked in 1973, "Some of us tried very hard to get good science done, and there was much bloodshed along the way." In retrospect, they succeeded admirably. At the fourth annual Lunar Science Conference, held in Houston in March 1973, leading Moon scientists adopted a motion that stated that despite years of conflict—characterized as "awkward moments" in the motion—the end results of Apollo research had "already revolutionized ideas of the solar system's evolution."

By the early part of the 20th century, ground-based astronomers had determined most of the physical features of the Moon. These astronomers had been quietly working for centuries to create the map that the scientists of the Apollo program used as a starting point for their explorations. This map's basic contours had been set down beginning in the 17th century and modified thereafter; it reflected the concerns, institutions, and knowledge of Western civilization during that time. Yet many questions remained unanswered. Key among them: What was the Moon's origin? How did the Moon form—at the time of the solar system's formation, or by capture at a later point? Did the Moon have a rocky or a dusty surface? Would the Moon's crust support any weight or would it swallow up anything that landed on it? How were the craters formed, via meteor strike or volcanic or some other type of activity? What was the structure of the Moon's interior? Did life ever exist in any form on the Moon? A 1969 assessment of lunar science undertaken prior to Apollo framed the key points of needed research, among them volcanism and meteor impacts, lunar stratigraphy, the Moon's origins and evolution, its surface and interior temperatures, and its geological formations.

Mistaken ideas about the Moon had lingered in parts of the astronomy community for years. One idea commonly held before Apollo suggested that life might exist on the Moon in the form of hardy microorganisms that could infect humans, and thus NASA quarantined the first astronauts returning from the Moon just in case they might return carrying some lunar bacteria or virus. The possibility of more complex life forms also held sway in certain circles. For example, a 1959 book with the ironic title *Strange World of the Moon* speculated on observed changes on the lunar surface:

Seasonal changes there certainly were: some markings darken,
others become paler, expand or contract, with a variation of hue,
in the course of a lunar day; nor are these changes symmetrical
as between evening and morning. The temperatures of the topsoil
are in step with the phase. Most of the seasonal changes, on the
other hand, lag behind the Sun two or three days, keeping pace
with the subsoil temperatures. This shows that they do not depend
on superficial alterations but have their seat some way below
the ground. Such indeed would be the behaviour of vegetation
sending long taproots down into the gas marsh.

Many astronomers investigating the Moon during the early years of the space age were also willing to consider the possibilities of a Moon not fully dead and alien. British astronomer Patrick Moore, before his knighthood, suggested that there might be vegetation in the nearside crater Aristarchus, where changing bands of color might signal the possibility of life hanging on near gaseous eruptions from underground. Indeed, many wanted to believe that the Moon was Earth's alter ego and therefore ascribed to it the presence of life.

Other theories about the Moon that later proved false also abounded. One of the most interesting was originated by Cornell University astronomer Thomas Gold, who insisted that the lunar surface had been eroded over the eons and that fine particles of dust had migrated from the lunar highlands to the mares. Dust covered the lunar surface, he believed, perhaps several miles deep, and anything that might land there would sink into a deep, dry quicksand, never to be heard from again. Few subscribed to this theory, but if true it meant that Apollo would be a fool's errand. Only after robotic explorers had investigated the matter as part of NASA's Surveyor program, which ran from 1966 to 1968, did Gold reluctantly relent in his commitment to this theory.

Finding evidence about the geological origins and development of this orbiting body became a central element of the Apollo science program. Their key questions—"How old is the Moon, how was it formed, and what is its composition?"—required the expertise of geologists rather than biologists, physicists, or other types of scientists, and therefore geologists led the science teams. NASA's senior partner in lunar science became the US Geological Survey (USGS), which dedicated considerable resources to the program, and USGS geologists such as Don E. Wilhelms, Michael Carr, and Eugene Shoemaker made important discoveries about the Moon.

In 1964, Homer E. Newell, NASA associate administrator for space science, convened the TYCHO group of scientists to assess the state of knowledge about the Moon, in one of several such studies the agency undertook. To point up the uncertain nature of lunar knowledge, the group noted in its 1965 report: "In considering the lunar surface it must be realized that up to the present all data available concerning the lunar surface was obtained through remote sensing." The study group noted that optical telescopes, for all of their capabilities, could not substitute for actual fieldwork. Until spacecraft and astronauts visited the surface, many questions would remain unanswered.

Executing the science of the Apollo program involved three hugely significant efforts: landing site selection, instrument and experiment selection, and training astronauts for scientific fieldwork. The first of these tasks was arguably the most important. While scientists had been planning since 1962 for investigations during each landing, they also got involved at the beginning with the Apollo Site Selection Board, established at NASA headquarters in August 1965. This board served as the primary vehicle for determining where each mission to the Moon would land, and thus the nature of scientific observation and experimentation that would be permitted there. It was always a contentious but necessary activity, and it successfully reached consensus on a range of geologically interesting landing sites.

The second major topic of debate was the definition of scientific experiments on the lunar surface. Ongoing debates about the size and mass of proposed experiments, as well as their power requirements, roiled mission planning efforts throughout the mid-1960s. The scientists agreed that the first investigations should focus on geology (especially sample collection), geochemistry, and geophysics. They also agreed that the early landings should concentrate on returning as many diverse lunar rock and soil samples as was feasible, as well as deployment of long-lasting surface instruments and geological exploration of the immediate landing areas by each crew. These tasks could be expanded later to include surveys of the whole Moon and detailed studies of specific sites in the equatorial belt.

Scientists would eventually place more than 50 experiments on the various Apollo missions, and in the case of the last landing mission, one of their own, Harrison Schmitt, undertook fieldwork on the Moon. The science packages deployed on the Moon included many experiments that together yielded more than 10,000 scientific papers and a major reinterpretation of the Moon's origins and evolution.

Finally, the scientific community worked with the Apollo astronauts to prepare them for geological fieldwork on the Moon. Never happy that academically trained geologists were rare in the astronaut corps—the only moonwalker was Schmitt, who had earned a PhD in geology from Harvard

This extraordinary lunarama, taken in December 1972 during Apollo 17, shows astronaut Harrison H. Schmitt's lunar rover at the Taurus-Littrow landing site. Buzz Aldrin called the Moon "Magnificent Desolation," an understatement that suggests better than any other the reason that the United States stopped going to the Moon: humans found nothing we wanted there. (NASA image no. AS17-137-21011)

University, while other geologists, such as Brian O'Leary, resigned from the astronaut program—they nonetheless worked hard to ensure that the flight crews had the knowledge and experience necessary to undertake useful work on the lunar surface. To a surprising degree, they succeeded.

Most scientists would probably agree with Schmitt, who commented in 1975 that "the Moon moves through space as an ancient text, related to the history of the Earth only through the interpretations of our minds, and, as the modern archive of our sun, recording in its soils much of imme-diate importance to man's future well-being." As reported in *Science* in

1973, "Man's knowledge of the moon has been dramatically transformed during the brief 3½ years between the first and last Apollo landing." Through a laborious polling of lunar scientists in the mid-1990s, the staff of the Curator for Planetary Materials Office at the Johnson Space Center compiled a list of the top 10 scientific discoveries made as a result of the Apollo expeditions to the Moon. Collectively, they described the current state of knowledge about this fascinating astronomical artifact, listing the following findings:

1. The Moon is not a primordial object; it is an evolved terrestrial planet with internal zoning similar to that of Earth. Before Apollo, the state of the Moon was a subject of almost unlimited speculation. We now know that the Moon is made of rocky material that has been variously melted, erupted through volcanoes, and crushed by meteorite impacts. . . .

2. The Moon is ancient and still preserves an early history (the first billion years) that must be common to all terrestrial planets. The extensive record of meteorite craters on the Moon, when calibrated using absolute ages of rock samples, provides a key for unraveling time scales for the geologic evolution of Mercury, Venus, and Mars based on their individual crater records. . . .

3. The youngest Moon rocks are virtually as old as the oldest Earth rocks. The earliest processes and events that probably affected both planetary bodies can now only be found on the Moon. Moon rock ages range from about 3.2 billion years in the maria (dark, low basins) to nearly 4.6 billion years in the terrae (light, rugged highlands). . . .

4. The Moon and Earth are genetically related and formed from different proportions of a common reservoir of materials. . . .

5. The Moon is lifeless; it contains no living organisms, fossils, or native organic compounds. Extensive testing revealed no evidence for life, past or present, among the lunar samples. . . .

6. All Moon rocks originated through high-temperature processes with little or no involvement with water. They are roughly divisible into three types: basalts, anorthosites, and breccias. . . .

7. Early in its history, the Moon was melted to great depths to form a "magma ocean." The lunar highlands contain the remnants of

early, low-density rocks that floated to the surface of the magma ocean. The lunar highlands were formed about 4.4–4.6 billion years ago by flotation of an early, feldspar-rich crust. . . .

8. The lunar magma ocean was followed by a series of huge asteroid impacts that created basins that were later filled by lava flows. . . .

9. The Moon is slightly asymmetrical in bulk form, possibly as a consequence of its evolution under Earth's gravitational influence. Its crust is thicker on the far side, while most volcanic basins—and unusual mass concentrations—occur on the near side. . . .

10. The surface of the Moon is covered by a rubble pile of rock fragments and dust, called the lunar regolith, that contains a unique radiation history of the Sun, which is of importance to understanding climate changes on Earth.

At the same time, the quest for knowledge about the Moon continues. Since the Moon landings, more than 60 research laboratories throughout the world have continued studies on the samples the Apollo missions collected. Many analytical technologies, including some that did not exist from 1969 to 1972, when the Apollo missions returned the lunar samples, have been applied by new generations of scientists.

Science, if it is to be useful, must not only generate and analyze new data but must also become a part of the body of knowledge available to society and incorporated into its milieu. Scientific knowledge of the Moon proliferated through a variety of forms over many years before Apollo, but the one that can be most easily tracked is information about the Moon and how it was presented in major textbooks and taught in informal and formal settings. Before study of the Apollo lunar rock and soil samples in the 1970s, controversy over lunar origins reigned among scientists, and competing schools of thought battled among themselves to establish prominence for their particular viewpoint in the textbooks; therefore determining the Moon's origins became the project's single most significant scientific objective. Three principal theories dominated:

1. Fission: the theory asserting that the Moon had split off Earth
2. Coaccretion: the theory holding that the Moon and Earth had formed at the same time from the Solar Nebula

3. Capture: the theory maintaining that the Moon had formed elsewhere and was subsequently drawn into orbit around Earth

Data supporting each of these theories had been generated and developed
to an amazingly fine point over time, but none of them, on its own, could
explain enough to convince a majority of planetary scientists.

The fission theory, first articulated in a concrete manner by George
Howard Darwin, son of Charles Darwin, in the 1870s, proposed that a
piece of mantle had detached from Earth while the planet was still young
and spinning fast. It formed into a ball, adherents argued, and entered
Earth orbit because of our planet's gravity. Debris left behind after the
separation then cratered the Moon and Earth. The fission theory enjoyed
general acceptance for more than a generation until physics undercut it:
scientists realized that the angular momentum of the Moon-Earth system had to be conserved.

Gradually, over the first half of the 20th century, the coaccretion and
capture theories emerged as alternatives to fission to help explain lunar
origins. Coaccretion argued that Earth and the Moon formed at the same
time, and from the same materials, that the solar system itself formed.
Capture held that a wandering asteroid was drawn in by the Earth's
gravity and became the Moon early in the formation of the solar system.
These theories had both adherents and detractors among the scientific
community, and no consensus existed on the subject.

Many scientists believed that Apollo science would lead to agreement
that one of these three major theories accurately described lunar origins.
Little did they know in the 1960s that new and detailed information
from the Apollo explorations would point toward a fourth option: impact
theory, which argues that the Moon formed from ejected material. The
"big whack" theory, as it is also known, explains well what was learned
about the geology and selenogony of the Moon during the Apollo program. In historian of science Stephen Brush's words, "According to this
hypothesis, the Earth was struck by a fairly large (perhaps Mars-size)
body, resulting in ejection of material that stayed in terrestrial orbit and
subsequently condensed to form the Moon." In 1984, at a lunar science
conference in Hawaii, the giant impact hypothesis was thoroughly discussed and gained at least provisional acceptance from most experts present, although many details remained to be worked out. As one participant

noted, this was "not because of any dramatic new development or infusion of data, but because the hypothesis was given serious and sustained attention for the first time. The resulting bandwagon has picked up speed (and some have hastened to jump aboard)."

Lunar scientist Paul D. Spudis explains further: "The giant-impact hypothesis appears to explain, or allow for, several fundamental relations—not just bulk composition, but also the orientation and evolution of the lunar orbit. . . . Part of the reason for this model's current popularity is doubtless because we know too little to rule it out: key factors such as the impactor's composition, the collision geometry, and the Moon's initial orbit are all undetermined." In the end, further research is required, for "as it turned out, neither the Apollo astronauts, the Luna vehicles, nor all the king's horses and all the king's men could assemble enough data to explain circumstances of the Moon's birth" to everyone's satisfaction.

To understand how lunar origins have been treated in science curricula and textbooks, I surveyed a succession of works on lunar and planetary science. In 1926, astronomers Henry Norris Russell, Raymond Smith Dugan, and John Quincy Stewart published a major work that discussed the origin of the Moon in relation to the fission hypothesis, the only explanation they offered for its origin. Their work went through several editions without significant alteration until the 1940s, without offering any theory of lunar origins beyond fission. Thereafter, other works tended to emphasize the three major theories—fission, coaccretion, and capture—as valid concepts that each claimed a loyal group of adherents within the scientific community, adding that none of them could be considered "settled science." With the rise of the Apollo program in the 1960s, as reported by William K. Hartmann in *Science*, "if our celebrated national lunar effort has produced any increased understanding, it ought to show up in the difference between 1967 lunar geology and the 1970 version. It does. There is a new tractability in the problems discussed. What has happened is not so much the cliché of 'astonishing growth of the field,' but rather the emergence of a healthy branch of science."

In 1974, Hartmann and Alastair Cameron, working independently, used lunar rock and soil samples returned from the Moon by Apollo astronauts to advance the theory that the Moon had been formed by debris from a massive collision with Earth about 4.6 billion years ago. A very large object, one as big as Mars (and later named Theia), had slammed

into Earth, they argued, and the Moon formed out of the ejected material. This theory fit neatly with the fact that the Moon does not have an
iron core, while Earth does: the debris blown out of both our planet and
Theia would have come from the bodies' iron-depleted rocky mantles.
Also lending credence to the theory was the comparative mean densi-
ties of Earth and the Moon. Earth has a mean density of 5.5 grams per
cubic centimeter, while the Moon's is only 3.3 grams per cubic centimeter
because it lacks iron. The Moon has the same oxygen isotope composition as Earth, whereas Mars rocks and meteorites from other parts of the
solar system have different oxygen isotope compositions. After the 1984
Hawaii conference, the impact theory promptly made its way into standard textbooks, first as only one theory among several. By 1990, however,
it had gained precedence as the preferred explanation in undergraduate
curricula.

Much has changed in terms of scientific knowledge about the Moon as
currently taught because of the Apollo effort, but broader public knowledge is another matter. Did the emerging consensus about the Moon's
origins change the manner in which most people lived their lives, and
did most nonscientists have any knowledge of it? In the years since the
first lunar landings, some Americans did begin to view the Moon more
prosaically than they had before, as more of a "big rock" than a mystery.
The Moon lost something of its wonder, its status as a possible haven for
other beings and its place as guardian of the nighttime. As one commentator noted in 2005:

> Like so many others, I have stood in line at the Smithsonian
> Museum in Washington to touch a piece of the moon, worn
> smooth under the pressure of millions of fingers. At this point we
> can say that the moon has been thoroughly demystified. We know
> what it is, what it is made of, and even know of its importance to
> the earth. When we gaze at the moon today, we do so with little
> of the awe and wonder of men thousands of years ago.

Even so, in the years since the last Apollo landings, the Moon has
retained some hold over humanity as an object of spiritual and cultural
fascination. It continues to inform our worldview, psyches, and even

religions to a degree that seems, at first glance, difficult to understand. As literary critic Randall H. Albright remarked in a discussion about William Blake, mysticism, reason, and science:

> We had people walking on the moon in 1969 through the 1970s. But did it demystify the moon? Not for me. And would Blake have approved? I highly doubt it. It did show that mathematical formulas, the scientific method, and technology can make a dream come true. And it did it have a positive ripple effect in some ways, like transistors, that helped "improve" human life, as well as a downside: a lot of garbage floating around earth's atmosphere, and diverting billions of dollars from social programs, national parks, and other ways to alleviate suffering here in the USA.

Less sophisticated assumptions abound concerning the place of the Moon in everyday life. In small ways, and some that are not so small, the Moon (especially in its full phase) remains mystical to the majority of Americans, linked in the public psyche to increased rates of crime, suicide, disasters, accidents, birth, and fertility, despite the lack of any empirical linkage. Numerous studies have established zero connection between the Moon and actions among humans. A 1988 review by Roger Culver, James Rotton, and I. W. Kelly in *Psychological Reports* that examined more than 100 studies on lunar effects concluded that all had failed to show a reliable and significant correlation between the full Moon, or any other phase, and the events on Earth often tied to them. Journalist John Roach reported, contrary to misconceptions from a variety of sources, that unusual behavior is no more likely to take place during a full Moon "than on any other night of the year, according to scientific reviews of the theory that the full moon alters the way humans and wildlife behave." Fortunately, most beliefs in lunacy—note the Moon connection—such as werewolves and vampires have dissipated in the past century.

Other lunar practices remain. Gardening by the Moon has been practiced for centuries, and despite the Apollo program's revelations, it remains popular today. Thousands of relatively intelligent individuals practice it for no reason based on scientific understanding, believing, in the words of one online guide, that doing so will "speed the germination of your seeds by working with the forces of nature. Plants respond to the same gravitational pull of tides that affect the oceans, which alternately stimulates root and leaf growth. Seeds sprout more quickly, plants grow vigorously and at an optimum rate, harvests are larger and they don't go to seed as fast." Some advocates of this type of agriculture also insist that plants' horoscopes determine how best to cultivate them. Whether one seeks to shore up such beliefs by pointing to gravity, electromagnetism, or some other force, no scientific evidence has confirmed that an extraterrestrial body more than 240,000 miles away exerts a controlling influence on humanity or other parts of the biosphere.

As another example, thousands of ordinary Americans attend full Moon meditations virtually every month, some with the explicit belief that something related to the Moon phase affects those meditating. As practitioner Alice A. Bailey commented, "Meditation is a potent method of service to humanity when the mind is used as a channel for the reception of the energies of light and love and the will-to-good and their direction into human consciousness. . . . In all aspects of our planetary life there are cycles in the ebb and flow of spiritual energies with which groups, as well as individuals, can consciously cooperate. One of the major energy cycles coincides with the phases of the moon, reaching its peak, its high tide, at the time of the full moon." Such ideas can be linked more explicitly to ancient worship of the Moon than to the demystified Moon of the modern age; full Moon meditations are a metaphysical practice, essentially a form of religion bearing no relationship to the modern scientific understanding of the Moon that Apollo made possible.

Each of these types of beliefs is aided in their perpetuation by four interrelated factors. First, and most important, these ideas about the Moon are ingrained in human consciousness by centuries of reinforcement. It is exceedingly difficult to overcome longstanding traditions with a few trips to the Moon by the Apollo astronauts. Second, the media reinforces such conceptions through repeated comments on them. Culver, Rotton, and Kelly have written of this: "With the constant media repetition of an association between the full moon and human behavior, it is not surprising that such beliefs are widespread in the general public." Anecdotal evidence of lunar effects is satisfactory for most people, and reporters tend to repeat anecdotes that others tell them.

Third, misconceptions about such things as the effects of the Moon's gravity on Earth support such ideas. Sociologists George O. Abell and

Bennett Greenspan undertook a study for *Skeptical Inquirer* in 1979 of birthrates during full Moons and found that the Moon actually exerts about as much gravitational pull on a pregnant woman (or anyone else) as the average insect does. Misconceptions about the physical nature of the Moon, they concluded, fuel widespread beliefs that bear no relationship to reality.

Fourth, communal reinforcement helps to garner support for these beliefs through repeated assertions. The process is independent of whether the claim has been properly researched or is supported by empirical data significant enough to warrant belief by reasonable people; a form of groupthink prevails in which only certain ideas are seen as valid, and competing positions might not be recognized. Through this process, untenable ideas are formulated, communicated, and passed on to succeeding generations.

These ideas might serve useful purposes at some level for some part of the community, but what does the persistence of myths about the Moon say about modern society? By replacing science with other, ritual forms of belief and communication about the Moon, those seeking to remystify it are drawn together in fellowship and commonality. The sharing of rituals helps to define a culture; people create and maintain realities through symbols found in or generated by art, science, religion, and literature. Sharing these realities is not just about communication; it also affirms social relations and creates a bulwark against the hard edges of the kind of science that Apollo uncovered. Something of a return to religion in place of science, it seeks meaning rather than data. Astronauts themselves have not been immune: Edgar Mitchell, lunar module pilot on Apollo 14, eventually devoted his time to mysticism.

In James Michener's 1982 novel *Space*, a character called Reverend Strabismus serves as a premodern foil to the scientific mindset emphasized in the story. Remorseless in his contesting of scientific accomplishment, he announces in one sermon:

Are you any happier because men claimed they walked on the Moon? Are your bills at the grocery store any lower? Are your children any better behaved? Are you pleased that these crazy doctors in London can make babies in test tubes? Or that abortionists are free to run rampant in this here land? . . . But I bring you release from all that. I tell you what you know in your hearts, there there's only one true way. Throw out these evil humanists.

Is the message here that for all of the scientific return of Apollo, many, perhaps most, of us prefer the comfort of folk wisdom to the scientific understanding of the Moon gained through Apollo investigation?

Recall the story of the three baseball umpires in chapter 1. Do we call strikes as they are, do we call them as we see them, or are they nothing until we call them? For people who choose to remystify the Moon, even after having been there themselves, the science brought to humanity by the Apollo program is viewed selectively, its interpretation dominated by perceptions that are not scientifically valid. In the decades since the Moon landings, has anything really changed? There are still those who, given the choice between the view of a mystical Moon and the knowledge that it is only a rocky sphere beyond Earth, without any magical influence over life on our planet, will always choose the former.

6

Apollo Imagery and
Vicarious Exploration

Everyone except the astronauts themselves participated in the voyages of exploration to the Moon by way of imagery captured more than 240,000 miles away. Some of the Apollo photographs have taken on iconic connotations in the decades since the last mission, and they have been replicated in many places and used in numerous striking ways. The power of the Apollo imagery served very specific needs for the United States at the time of the Moon landings—especially the need to enhance national prestige—and in the years afterward, the images have been mobilized mostly to buttress the same sort of perspectives. Apollo photos speak to the expansive manner in which the United States took its measure among the nations of the globe: by creating such powerful and unique images, the country gained stature in the eyes of the world. The imagery returned to Earth by the astronauts during Apollo offered (and still does) an archetypal statement of American ingenuity, technological virtuosity, national exceptionalism, and the power of the state to accomplish useful things.

There is a long history of photography in exploration expeditions. The ability to document, through visual means, the "conquest" of the poles, mountains, undersea sites, or any other area on the globe has proved critical to both explorers and colonial powers. For example, in 1871, when Ferdinand V. Hayden, head of the United States Geological and Geographical Survey of the Territories, led an expedition into the Yellowstone region of the northern Rockies, he took with him two artists, Henry W. Elliot and Thomas Moran, and a photographer, William H. Jackson, to document the expedition's findings. Jackson's images in particular became the stuff of legend, merging humanity and nature, wonder and the mundane, to capture the essence of the lands the survey had traversed. So important were the photographs that Hayden said of them before Congress after returning to Washington, DC:

They have done much, in the first place, to secure truthfulness in the representation of mountain and other scenery. Twenty years ago, no more than caricatures existed, as a general rule, of the leading features of overland exploration. Mountains were represented with angles of sixty degrees inclination, covered with great glaciers, and modeled upon the type of any other then the Rocky Mountains, the angular lines of a sandstone mesa, represented with all the peculiarities of volcanic upheaval, or of massive granite, or an ancient ruin with clean-cut, perfectly squared and joined masonry, that would be creditable to modern times. The truthful representations of photography render such careless work so apparent that it would not be tolerated at the present day.

Photography became critical to explorations of all the remaining frontiers, from the Rocky Mountain expeditions undertaken by John Wesley Powell and others, to travels to the sources of the Amazon River in South America and the Nile River in Africa, and journeys to the poles with Robert Peary and Richard Byrd. Photos documented explorers' accomplishments and were especially critical to international scientific endeavors such as the International Polar Years of 1882–83 and 1932–33, which cooperatively sought to obtain scientific and ethnographic data about the Arctic and Antarctic. They furthermore captured scientific data, which could be used to transform understanding of remote and exotic regions while also serving to cement the names and exploits of European and American explorers.

Photography proved itself useful in four key ways in scientific expeditions. First, the imagery was a documentary record of the expedition, and as such it offered scientific data that could be mined for understanding. Second, photos served the important task of demonstrating that an exploration actually had taken place. Challenges to the veracity of expeditionary claims are as old as the first voyages of human history. In the early 1860s, for example, Sir Richard Burton and John Hanning Speke both publicly sought the source of the Nile, with Burton questioning the findings of an expedition Speke had led to find it. Speke died of a gunshot wound while hunting in 1864 just before he was due to debate Burton on the matter in front of the British Royal Geographical Society. Some believed he had committed suicide to avoid the debate, although his friends insisted that it was an accident. Things might have been different had Speke's expedition been visually documented by photography.

Third, these explorations signaled a public message of commitment to progress and a prosperous future. Always, in every continent on which Western civilization gained a foothold, they served as the harbingers of economic activities that would ultimately exploit the lands brought under control. Sometimes that effort was brutal, and in many instances the conquest decimated the local population. Even in unpeopled regions such as the poles, the sense of progress, and all that it meant for commerce, industry, and exploitation of natural resources, reigned supreme. The photographic record of an expedition initially revealed how regions looked when they were first explored, and then each successive expedition's images would show new settlements and towns and eventually the alteration of the landscape itself. These changes often were specifically documented with a series of images shot from the same spot over periods of years. In each image, the viewer can perceive the "march of progress."

Finally, the sense of prestige and honor that accrues to those undertaking exploration has been bolstered by documentary imagery since the invention of the camera. Ernest Shackleton's attempt to reach the South Pole during his third expedition, for example, was documented in the stunning photography of Frank Hurley. The expedition failed, and while Hurley's photos did not offer much scientific data about Antarctica that could be analyzed, they did reveal the heroic efforts of Shackleton and his men to return safely home to England. Photos of John Wesley Powell standing before the Grand Canyon, Henry Morton Stanley seeking David Livingstone at Lake Tanganyika in Africa, and Robert E. Peary at the North Pole over time became similarly iconic, fixing the individuals and the expeditions in the minds of the public and serving, in the words of Stephen J. Pyne, as "an index of national prestige and power." Photography and eventually film would become the methods of choice to demonstrate such undertakings.

Apollo imagery served to accomplish the same four central tasks: providing scientific documentation, demonstrating the nature of the effort, modeling progress, and asserting prestige. Whether or not NASA intended it, its astronauts could not have done a better job of documenting their journeys had they been following a rigid script. In taking these photos, often with themselves as the centerpiece of the imagery, the astronauts personalized their exploits in the same way that Powell had done during his Colorado River excursions. The images placed a hero before the public, a gesture that served the national interest in ways both obvious and sublime.

On the first US spaceflight programs, astronauts could not take their own photos. Alan Shephard's and Gus Grissom's Mercury flights included only cockpit cameras that recorded the astronauts' reactions to the rigors of spaceflight. While NASA released these images to the public, they proved less inviting than the more free-form imagery of later missions. John Glenn took the first camera into space during the flight of Friendship 7 in February 1962 and took photographs of Earth's features and meteorological phenomena during his three-orbit mission. According to legend, Glenn purchased the camera in a drugstore before the flight. The actual story was a bit more complex. Robert Voas, a Mercury astronaut training official, assigned Glenn the task of taking pictures through the window and periscope with a handheld camera as part of an experiment to test human capabilities during orbit. He used a Minolta HiMatic 35mm camera, sometimes marketed as an Ansco Autoset, that had been modified to reduce its weight and to add a handle so that the left-handed Glenn could operate it. After the flight, he reflected: "It just seemed perfectly natural; rather than put the camera away, I just put it out in mid-air and let go of it." He did have some problems trying to change the standard and ultraviolet-sensitive film on which he shot, saying of one slippery roll, "Instead of clamping onto it, I batted it and it went sailing off around behind the instrument panel, and that was the last I saw of it."

Government officials had some political concerns about astronauts' taking photographs from space. In the midst of the Cold War, and with reconnaissance satellites starting to operate over the Soviet Union, NASA leaders wondered if unfriendly nations might view such photography as a form of American belligerence. Meanwhile national security policymakers sought to put into place procedures ensuring that imagery obtained by NASA astronauts was not released to the public until it underwent security analysis. Mercury project managers, for their part, worried that the small capsule could not accommodate a camera and its accouterments. There was no acceptable place to stow it. Nonetheless photography became a priority for each mission after Glenn's success on the first orbital flight. Faced with this emerging requirement, project scientists developed a series of experiments for the astronauts that included horizon definition, weather, terrain, and other specific types of photography.

NASA did not initially intend to use these images in publicity for the space program. For instance, scientist Max Peterson, at the MIT Instrumentation Laboratory, had asked the Mercury astronauts for photos that could be used "in determining the effectiveness with which the earth's sunlit limb could be used for navigational sighting during the terminal phase of advanced space missions." He found that "limited results of the MA-7 flight have shown, as expected, that the earth's limb viewed through a blue filter has a somewhat higher elevation than when viewed through a red filter. . . . The limb viewed through a blue filter is expected, however, to provide a better navigational reference because the blue limb appears more stable and is not as subject to interference effects from clouds and other atmospheric conditions as is the red limb." Both the experiments and the cameras became more sophisticated as time progressed.

NASA's intentions aside, photography of Earth from space and shots of the astronauts themselves proved to be highly significant for the space program. The public never seemed to tire of viewing photos of Earth from space and of the astronauts going about their responsibilities in orbit. More opportunity to provide such images came during the Gemini program in 1965–66, when two-astronaut crews flew 10 low-Earth-orbit missions and the activities they conducted became more complex. This period also saw the first coordinated effort to use such imagery to sell the public on the merits of space exploration. Richard Underwood, who worked in this arena for NASA in the 1960s, recalled that it was the Gemini IV mission in 1965, when Edward White undertook the first spacewalk and took photos in orbit, that gave the agency's leaders a new appreciation for the power of imagery. Robert Gilruth, director of the Manned Spacecraft Center, said, "We're looking at things that no human being had ever seen before, parts of Africa and other places. You can see what really goes on." Gilruth immediately embraced the use of cameras on all subsequent missions, telling Underwood, "From now on, your job is to work with the astronauts to be sure they bring back great photographs of the Earth and then eventually when we go to the Moon."

NASA moved as swiftly as possible to make photos from each mission available to the public. More than 650 Earth and weather images from the first three Gemini missions alone were swiftly distributed to the media. NASA gave the astronauts Hasselblad 550C medium-format cameras, assigning the crews specific sites on Earth where they should photograph "selected cities, rail, highways, harbors, rivers, lakes, illuminated night-side [sic] sites, ships and wakes." Overall, NASA astronauts took more than a million images on their Apollo spacecraft flights during the Moon-landing program. As Underwood told the astronauts:

You know, when you get back, you're going to be a national hero, you're going to get a parade in your home town and maybe a parade in Washington, and you're going to have dinner at the White House and you're going to talk to a joint session of Congress. All through this time, all these computers in Building 30 are going to be punching information into big thick books about what went on in this system, that system, and the other system and so on. The only people who are ever going to look at those books are probably guys going for a Ph.D. in aerospace engineering or history of this sort of thing, and they've got a billion pieces of data on thousands and thousands of these books. That's their only value. But those photographs, if you get great photos, they'll live forever. Your key to immortality is in the quality of the photographs and nothing else.

Underwood added that sometimes the astronauts disagreed with his reasoning but would then think further about it and come back to tell him he was probably right. Some of them became enthusiastic pho-tographers. Of the imagery they took, only a small percentage actually found its way into broad public distribution. No shots were suppressed or prohibited from release, but only the best of the lot found a public audience. Ineffectively centered, dully lighted, poorly conceived images were relegated to the archives. For instance, the famous *Earthrise* photo-graph (page 114) taken from Apollo 8 is not the only image of Earth rising over the lunar horizon; it is the most striking and memorable of several similar images that the mission's astronauts took with two cameras (one loaded with black-and-white film and the other with color film), and thus it was the shot chosen for broad dissemination. Such has long been the standard practice of photographers, offering for public scrutiny what they consider the best of their imagery while throwing away those deemed of lesser value.

Much of the power of the imagery emerging from Apollo resulted from the juxtaposition of the terrestrial with the extraterrestrial, from the alien with the familiar, in the same frame. The Apollo images represent what Roland Barthes referred to as the duality of existence, or "the co-presence of two discontinuous elements, heterogeneous in that they did not belong to the same world."

At the same time, NASA's astronaut photography fell into a genre well defined by the 1960s era of social realism and reform. In so doing, it followed a model developed by early-20th-century reform movements, as outlined by Maren Stange:

Photography, uniquely documentary and mass reproducible, became particularly useful to reformers intent on communicating a worldview that stressed their expertise and organization. In order to assert more or less explicitly that their images presented viewers with truth, reformers relied on the photograph's status as *index*—that is, as a symbol for fulfilling its representative function "by virtue of a character which it could not have if its object did not exist," in a standard semiotic definition.

The documentary nature of photography, apart from the reform dimension, served NASA well in establishing the reality of the Apollo episodes. At the same time, the most impressive, expressive, and salient images of Apollo adopted the iconography of the sublime, inspiring awe and wonder. Their dramatic light and dark features, the colors of the human-made objects in images set upon the dull gray palette of the lunar surface, made for dynamic compositions suggesting a sense of vastness, infinity, and greatness. Always these images harked back to the photography of earlier exploring expeditions. While the astronauts did not model their efforts consciously on those earlier works, they became acutely aware of the visual power of their imagery and their ability to similarly inspire emotions and excitement.

Five Apollo images have taken on especially iconic connotations in the years since the last mission:

Buzz Aldrin and the American flag on the lunar surface, Apollo 11
Full image of Aldrin from Apollo 11
Bootprint on the Moon, Apollo 11
Earthrise, Apollo 8
Whole Earth, Apollo 17

While other imagery might be useful for a variety of purposes, these five photos best demonstrate the specific national sense of pride in the Apollo accomplishment. The four themes of expeditionary photography threaded their way through the reception of each one. They documented in graphic detail the scientific accomplishments of the missions to the Moon. They demonstrated too that the missions had actually taken place, effectively allowing the public to accompany the astronauts to the Moon. They signaled a public message of progress and a prosperous future, and they presented, with clarity and alacrity, a national sense of prestige and honor as they offered the missions' stunning feats to a world that might otherwise have been skeptical. Notable in the list above is the fact that—despite the increasing sophistication of the photographic equipment available to the astronauts, the ever-evolving complexity of the photographic regimen, and the ramped-up training of astronauts in photography as Apollo progressed—the images that have entered the realm of public iconography are mostly those from the early lunar expeditions, which speak to a larger lifeworld of hope, and power, and progress.

The first of the iconic images of Apollo was taken in 1969, when Neil Armstrong photographed Buzz Aldrin at the US flag (page xiv) after they had planted it on the lunar surface. This image circled the globe and was replicated in a variety of settings. It remains an important element in the cultural derivation of Apollo, the Moon landing, and the personification of American prestige. The stark duality of the human-made world impinging on the natural, untouched, pristine, otherworldly lunar terrain obviously has saliency beyond simply inspiring a sense of the sublime. The prominence of the flag, its colors and forms, signals the triumph of the American spirit, the success of the endeavor. Nothing is a more powerful image of national identity. It signaled to the nation that

NASA and its empowered bureaucracy had accomplished the seemingly impossible, and it conveyed the same sentiments to the world's population, which celebrated and recoiled from the American achievement at the same time. At the time of the Apollo 11 landing, Mission Control in Houston flashed the words of President Kennedy announcing the Apollo commitment on its big screen, followed by these words: "TASK ACCOMPLISHED, July 1969." Probably there could be no greater understatement made, but NASA ensured that everyone knew exactly who had completed this seemingly impossible task. Glory, honor, and prestige converged in recognition of the accomplishment of landing an American on the Moon.

The unfurling of the flag was modestly controversial at first. In 1965, the Gemini IV crew of James McDivitt and Edward White were the first to decorate their spacesuits by adding American flags on their shoulders. They purchased the patches themselves, but they set a standard for the future crews, who also adopted flag patches on their suits. No one questioned that development, either in NASA or outside it, but when it came time to consider planting a flag on the Moon, national prestige butted up against international sensitivities. While raising a flag on the lunar surface might have seemed merely a symbolic gesture to most Americans—an expression of triumph and uniqueness—it raised the hackles of some allied nations, whose leaders urged a broader perspective. After Richard Nixon sounded an internationalist note in his inaugural address ("As we explore the reaches of space, let us go to the new worlds together—not as new worlds to be conquered, but as a new adventure to be shared"), some argued for the raising of the flag of the United Nations rather than the United States. NASA's Committee on Symbolic Activities for the First Lunar Landing considered this possibility but eventually decided that the US flag would be the only one unfurled on the Moon. The plaque saying, "We came in peace for all mankind" would commemorate the larger international perspective. While the original plaque design also featured a US flag, NASA officials decided to make its focus more global by substituting artwork depicting Earth's Eastern and Western Hemispheres. Apollo 11 also carried small flags from all member countries of the United Nations, as well as a few other nations, taking them down to the lunar surface and bringing them back to Earth as gifts to various heads of state.

The flag controversy has arisen many times since, mostly recently (as of this writing) in August 2018, when the feature film *First Man*, about the life of Neil Armstrong, premiered to strong acclaim—except for one part, in which director Damien Chazelle explicitly omitted the flag scene from Apollo 11. Actor Ryan Gosling, who portrayed Armstrong in the film, responded to questions about this decision by saying that Chazelle believed the achievement "transcended countries and borders." He added, "I think this was widely regarded in the end as a human achievement [and] that's how we chose to view it." This bothered some Americans on the political right, who believed that the United States deserved full credit for the lunar landing and that anything less was unpatriotic. Senator Marco Rubio (R-FL) tweeted his disagreement: "This is total lunacy. And a disservice at a time when our people need reminders of what we can achieve when we work together. The American people paid for that mission, on rockets built by Americans, with American technology & carrying American astronauts. It wasn't a UN mission." Aldrin, the astronaut who raised the flag at Apollo 11 alongside Armstrong, agreed. He tweeted, along with pictures of the flag planting in 1969, "#proudtobeanamerican." The indignation continued thereafter, but it did not dampen enthusiasm for the film. Did the internationalization of the Moon landing make a difference in the film's success? Probably not, but the controversy over this omission suggests the power of the iconic flag-planting image.

After Apollo 11, Congress weighed in to forestall any other NASA efforts to broaden lunar flag-raisings beyond the US flag. It passed a law on November 4, 1969—too late for Apollo 11 but intended to be ironclad guidance for future flights—saying that "the flag of the United States, and no other flag, shall be implanted or otherwise placed on the surface of the moon, or on the surface of any planet, by members of the crew of any spacecraft . . . as part of any mission . . . the funds for which are provided entirely by the Government of the United States." It was the final word on a statement of pride and prestige that was meant to be American, and only American.

Armstrong and Aldrin's deployment of the flag together on July 20, 1969, was displayed to a worldwide audience on live television. (As noted earlier, they had difficulty placing the pole in the regolith and extending the horizontal telescoping rod on which the flag hung, and thus the flag looked as if it were blowing in a breeze—something the Moon-landing deniers have long used as "evidence" of a NASA conspiracy to fake the Moon landings despite the obvious explanation—in an effect mimicked by later crews.) The flag-raising, not only the most-watched public event of the mission, may also have been its most significant. The still image has found publication in all manner of newspapers, magazines, and books since that time. It was a deeply moving experience for many people. Aldrin would later recall that he felt an "almost mystical unification of all people in the world at that moment."

Not surprisingly, this imagery supported a broad narrative of "America-first" globalism. While the United States might have come in peace for all mankind, those words masked a beguiling nationalism. In images such as Aldrin at the flag, the United States restated, in a way that might never have been so powerful otherwise, that its frontier truly extended into outer space, and recast its interests as the world's interests. The Apollo imagery, therefore, assisted in constructing a particular brand of global hegemony firmly rooted in the concept of American leadership in the world.

The ways in which the image of Aldrin at the flag has found use are extraordinary, attesting to the power of its symbolism. An artist's rendering appeared on the cover of *Time* magazine on July 25, 1969. It appeared as artists' conceptions on a succession of postage stamps during the later 1960s and around various landing anniversaries. Numerous works of fine art have depicted it in various ways. Perhaps the most famous and striking is *Moonwalk*, Andy Warhol's 1987 silkscreen of an image of Aldrin standing on the Moon with a flag at his side, a print whose neon colors still conjure up nostalgia for the '60s psychedelic era as well as the buttoned-down discipline of the Moon-landing era. Interestingly, Warhol made the image his own, in a way similar to Jimi Hendrix's reinterpretation and claiming of "The Star-Spangled Banner" at Woodstock, by including the initials A. W. in neon colors on Aldrin's visor in place of Armstrong's reflection. Thus the artist took ownership of the scene in a way directly reminiscent of the flag-planting itself. By turning a symbol of American patriotism on its head, Warhol, like Hendrix, offered a meaningful critique of the society in which he lived. Warhol's sudden death left this series of prints unfinished, and one may ponder where it might have gone had he been able to complete it, but in some ways it is among his most striking visual work.

A dizzying and confounding array of consumer products related to an astronaut placing the national flag on the Moon has been created to separate undiscriminating shoppers from their money. All manner of figurines and plastic statues have come and gone from the market. One of the most outlandish is the "Cow Has Landed on the Moon" collectible. Part of a larger set of collectibles featuring Holstein cows—the Parade of Cows set, presently available at the NASA Visitors' Center in Houston—one such figure shows a cow planting the Texas state flag on the Moon with this caption at the base: "Houston, We Have Landed." This item capitalized on an ingenious public art effort in the United States. Beginning in Chicago and New York in 1999 and 2000 and moving to other cities thereafter, local artists were challenged to create unique statues of cows in various settings and clothing and were encouraged to offer their own interpretation of the cow as an art object. Not intended as high art by any means, this public art effort offered humor and sometimes insight. (The collectibles started a craze in many quarters and can be seen at cowparade.com, described by its owners as "your virtual 'moo-seum' for all of the CowParades, past, present and future.")

Perhaps the most startling and recognizable use of this imagery was the cable television MTV network's logo. In 1981, MTV began operations and was searching for a new look to promote its then-revolutionary idea of a 24-hour music channel. Working with Manhattan Design Inc. to obtain the big block M and graffiti TV logo, designers then placed it on the US flag in the famous Apollo photograph. They chose this image in part because it was so recognizable and in part because it required no licensing fee, since it was a public-domain photo. The irony of MTV planting its flag on the Moon for the entire world to see was not lost on those designing the imagery, and the "Moonman" and the MTV flag soon became a regular network feature, used to introduce its "Top of the Hour" programming. The Moonman found itself incorporated into video advertisements and into the statuette for the MTV Video Music Awards, presented each year since 1984.

A work in the National Air and Space Museum depicts an astronaut on the Moon with the flag. *Space Mural—A Cosmic View* was painted by Robert T. McCall in 1976 and has been a favorite of visitors ever since. It is a popular point for family photography, and I have even noticed children bowing to the astronaut depicted on the lunar surface on occasion.

The MTV Video Music Awards has used some version of an astronaut on the Moon in its trophies since the awards show began in 1984. This "Moonman" statuette, long one of the most recognized trophies in the world, was given by MTV Networks to the Smithsonian National Air and Space Museum in 2007. (NASM image no. T20080004001cp08)

What does it say when families, many from other nations, pose before the astronaut with the US flag for their family group shots? An answer to that question is suggested by a critic's description of the photo upon which the painting is based: "A photo of Aldrin, standing next to the crisp and bright flag, captures the ideals of a generation and marks the end of one aspect of the Cold War—the space race between the Soviet Union and the United States." This was the conclusion intended by NASA and the nation's leaders when they conceived of the flag-raising ritual before the flight of Apollo 11. It succeeded beyond their expectations.

Perhaps the most famous of all of the images from Apollo, however, is the 1969 photograph of Aldrin facing the camera on Moon, his helmet visor showing a reflection of Armstrong, who took the photo (page xxi). Also visible is one footpad of the lunar module *Eagle*. This shot graced the front page of newspapers around the world after the Moon landing and served as the cover of *Life*, *National Geographic*, and a host of other magazines in the weeks following.

It captures well the iconic aspects of the Apollo program. It shows the astronaut fully encased in his armorlike suit, battling an impossible environment and other hazards that threaten his life. It emphasizes the transcendent power of the Apollo program, the literal otherworldliness of a human on the Moon juxtaposed with the stark alienness of the lunar landscape. It also serves to assign supreme masculinity to the endeavor: the astronauts, like medieval knights, overcame forces that seemed insurmountable.

Like the other images remembered by the public long after Apollo, the photo of Aldrin facing the camera on the Moon has been used in a variety of settings ranging from fine art to advertising to NASA publicity to journalism. Aldrin recalled the experience of posing for the picture in 1999:

As the most evocative image from our landing on the Moon, I would choose the picture that the world has picked: The one that shows the reflection in my visor. The picture taken just before that shows me walking toward the lunar module. Neil took the first picture, a profile, then he said, "Stop and turn." I stopped, turned and looked at him—and immediately he took it. So that was a spontaneous picture, not at all posed. My hands wouldn't have been that way, with the arm kind of crooked.

It is almost inescapable as a representative image from Apollo, and it has routinely graced nearly every book and article written about the subject in the 50 years after it was taken.

Almost as famous as the two Aldrin images is the photo showing a bootprint on the lunar surface, which Aldrin took during the Apollo 11 mission. Aldrin had the task of photographing the lunar soil for scientific purposes, but he also recognized the importance of the footprint on the Moon for the public—it would be, perhaps, the ultimate in vicarious exploration. In his framing, the natural features of the Moon's surface are seen around the human imprint that proves we were there. As Aldrin recalled of capturing the shot:

First I took a picture of the surface of the Moon; then I put my foot on it. You see the same little grains [of moondust] here and there—but at a slightly different angle, because I had to move. Then I said to myself, "Now wait a minute, that's a lonely-looking bootprint; let me take a picture of the boot." So for the next picture

I made a new bootprint, then moved my foot an inch or two away—so that you see a picture of my overshoe, poised above the boot print.

What does this image convey? Many clever conceptions have been suggested:

I shall return
A lasting first impression
Remove boots before entering
Take nothing but pictures, leave nothing but footprints
Time to make more
Next time we return to stay
Down payment on an outpost
When do I get to go?
Who might tread over this bootprint?
What will aliens think if they see this footprint?

What became obvious immediately after people on Earth saw the photo was that the bootprint would remain on the lunar surface indefinitely. Because of Earth's geological and climatic activity, footprints and other such indicators of fauna generally do not remain long. Not so on the Moon. Since it has no atmosphere, no weather, no life, and virtually no geological activity, there is little to erase the iconic bootprints that Armstrong, Aldrin, and the other Moonwalkers left there. In the near future, only careless human and robotic visitors might be able to destroy the prints, so important in history. This is such an important issue that several preservationists have emerged in the first decades of the 21st century to advocate protection of the footprint locations as heritage sites. As an example, Michelle O'Hanlon has established For All Moonkind, Inc., to pursue a strategy for legal protection of the bootprints. Others have pursued this goal through other avenues, engaging NASA and the United Nations Committee on the Peaceful Uses of Outer Space (COPUOS). Nothing has happened yet, but it might well in the future.

Without question, another of the critically significant images from Apollo is the photograph known as *Earthrise*, taken by the Apollo 8 crew on Christmas Eve 1968. The shot was captured while the spacecraft

Earthrise, one of the most powerful and iconic images from the Apollo program, was taken by astronaut Bill Anders in December 1968 during the Apollo 8 mission. This view of the rising Earth greeted the Apollo 8 astronauts as they emerged from behind the Moon after the first lunar orbit. Often used as a symbol of the planet's fragility, the photo juxtaposes the gray, lifeless Moon in the foreground with the blue-and-white Earth, teeming with life, hanging in the blackness of space. (NASA image no. 68-HC-870)

was at 110 degrees east longitude and the horizon, about 350 statute miles from the spacecraft, was near the eastern limb of the Moon as viewed from Earth. On our planet, hanging in the dark void of space, the sunset terminator is seen crossing Africa. Antarctica shows up well as a white swatch; North and South America are almost hidden under clouds. Earth, awash in color, makes a stark contrast to the grays and browns of the lunar surface and the blackness of space. This photograph had the most profound effect on public consciousness of any image captured the lunar surface, coming to symbolize a new era of environmental consciousness and concern for the ultimate fate of the home planet.

The Apollo 8 crew itself was fascinated by this image. Astronaut Bill Anders recalled the experience in a 1979 interview:

We had been in orbit for three orbits, and for some reason had not seen the Earth . . . we had been trained to look at the moon, we hadn't been trained to look back at the Earth . . . and yet when I looked up and saw the Earth coming up on this very stark, beat-up lunar horizon, and Earth that was the only color that we could see, uh, a very fragile looking Earth, a very delicate looking Earth, I was immediately almost overcome of the thought here we came all this way to the moon, and yet the most significant thing we're seeing is our own home planet, the Earth.

Frank Borman added his own thoughts about the event:

We were the first humans to see the world in its majestic totality, an intensely emotional experience for each of us. We said nothing to each other, but I was sure our thoughts were identical—of our families on that spinning globe. And maybe we shared another thought I had . . . *This must be what God sees.* . . . I happened to glance out of one of the still-clear windows just at the moment the earth appeared over the lunar horizon. It was the most beautiful, heart-catching sight of my life, one that sent a torrent of nostalgia, of sheer home-sickness, surging through me. It was the only thing in space that had any color to it.

At a fundamental level, *Earthrise* became the ultimate reconnaissance photograph of Earth taken from afar. It has been reproduced in all sorts of settings and places since it first entered the public's consciousness in December 1968. James Lovelock, godfather of the Gaia hypothesis, offered this perspective on the image and its place in the iconography of Apollo:

Can there have been any more inspiring vision this century than that of the Earth from space? We saw for the first time what a gem of a planet we live on. The astronauts who saw the whole Earth from Apollo 8 gave us an icon that has become as powerful as the scimitar or the cross. In the years leading up to this mission in 1968, I had worked with the American National Aeronautics and Space Administration (NASA) and had seen behind the scenes.

The meaning of that cloud-speckled ocean-blue sphere was made real to me by their newly won scientific information about the Earth and its sibling planets Mars and Venus. Suddenly, as a revelation, I saw Earth as a living planet. The quest to know and understand our planet as one that behaves like something alive, and which has kept a home for us, has been the Grail that has beckoned me ever since.

No less significant than *Earthrise* is the *Whole Earth* image captured from Apollo 17 on December 7, 1972, also one of the most widely distributed photographic images in existence. This image of a fully lit Earth is the best among many taken by Apollo astronauts, as the Apollo 17 crew had the Sun behind them when they took the photograph. Sometimes called *The Blue Marble*, this image, taken during the translunar coast en route to the Moon, shows the northern coast of Africa and a sliver of the Mediterranean in the north and extends southward far enough to capture some of the Antarctic ice cap. Heavy cloud cover is seen in the Southern Hemisphere, but most of the coastline of Africa is clearly visible, as are the Arabian Peninsula, Madagascar, and portions of the Asian mainland.

As early as 1966, environmental activist Stewart Brand had begun a campaign for NASA to release an image of the whole Earth in space. He even made up buttons that asked, "Why haven't we seen a photograph of the Whole Earth yet?," selling them on college campuses and mailing them to prominent scientists, futurists, and legislators, but not until Apollo 17 would *Whole Earth* become a reality. As Brand recalled of the early days of his campaign:

I turned on my blanket there on the gravel rooftop and looked clear around, it was indeed a circle, a mandala—a nice, finite, entire, low-altitude view of the Earth. . . . I just sat all afternoon and tried to think of how we could possibly get a photograph of the whole Earth—that is, of the planet from space. I was a big fan of NASA and of [the] then ten years of space exploration that had gone up to that point, and there we were in 1966, having seen a lot of the moon and a lot of hunks of the Earth, but never the complete mandala. . . . It was a bit odd that for ten years, with all

One of the most widely recognized images from Apollo, *Whole Earth* was taken by the Apollo 17 crew as they traveled toward the Moon in December 1972. Perhaps the most significant use of the photo has been its deployment by the modern environmental movement to depict Earth as a fragile orb of life in the void of space. (NASA image no. AS17-148-22727)

the photographic apparatus in the world, we hadn't turned the cameras that 180 degrees to look back.

This story has been told and retold in various ways, with some authors suggesting that Brand once alleged a NASA cover-up of secret photographs, although his statements do not reflect this idea.

To capture this iconic image, the Apollo 17 astronaut-photographers used a 70-millimeter Hasselblad camera with an 80-millimeter lens. It has been virtually impossible to figure out which person among the crew—Eugene Cernan, Ronald Evans, or Harrison Schmitt—actually took the photo, because they all took many shots with the Hasselblad during the mission. Many credit Schmitt with snapping the photo, but that cannot be determined for certain.

Apollo 17 was the last lunar landing mission for the United States. Since no humans have made the trip to the Moon since, it was the last occasion on which such an image could have been taken. The photo has been used repeatedly since that time, especially in the cause of Earth's ecology and its preservation. Brand put the photograph on the cover of his *Whole Earth Catalog*, which he had been publishing since 1968 but now brought out in an oversize paperback edition with the world on its cover. This image, and the other stunning photographs of Earth taken from space, inspired a reconsideration of our place in the universe. It has become the key image deployed by environmental activists, politicians, and scientists during the annual Earth Day celebrations (which began in 1970). It suggests an object lesson that Earth is a small, vulnerable, lonely, and fragile body teeming with life in a dull and lifeless black void, and it implies that humanity, although self-regulating and ancient, has proved to be a threat to this place and that Earth requires human protection.

The *Whole Earth* image, as well as the earlier *Earthrise* photograph, prompted the people of the world to view Planet Earth in a new way. Writer Archibald MacLeish summed up the feelings of many when he wrote during the Apollo years, "To see the Earth as it truly is, small and blue and beautiful in that eternal silence where it floats, is to see ourselves as riders on the Earth together, brothers on that bright loveliness in the eternal cold—brothers who know now that they are truly brothers." The modern environmental movement was galvanized in part by the photo's new, remote, holistic viewpoint on the planet, which implied a need to protect both it and the life that it supports.

These five key images from the Moon landings have been replicated in many places and used in numerous striking ways. Most important, however, the powerful Apollo imagery answered specific needs for the United States at the time of the Moon landings, and it has largely been mobilized to bolster the stature of the nation in the years since the last flight and to fulfill the dominant narrative about the Apollo program as a representative American success. In the process, it has served as an exemplar of a grand visionary concept for human exploration. Given this observation, Apollo has been celebrated as an investment in technology, science, and knowledge that would enable humanity—or at least

Americans—to do more than just dip its toes in the cosmic ocean, to become a truly spacefaring people.

The use of this imagery as evidence of Apollo's powerful incantation for a grand national vision of the future sparks considerable nostalgia for the exciting experience of astronauts walking on the Moon. While there were setbacks, in this interpretation the experience of the 11 years of Apollo between 1961 and 1972 contained more triumph and tragedy, more heroic sacrifice, more strenuous effort than many wars and certainly the years of space exploration since. Many remember Apollo as an effort wrought with high drama and excitement. More important, Apollo was a government program that actually succeeded. Twelve American astronauts did indeed land on the Moon and return safely to Earth, achieving the first of those landings "before this decade is out," as John F. Kennedy had directed.

At the same time, this imagery has been juxtaposed with imagery and ideas from American social upheaval, defeat in Vietnam, and Great Society emphases of the 1960s to point up criticism of the program from both the left and the right of the political spectrum depending on the ideals emphasized. Finally, some of these same images have been exploited by Moon-landing deniers to question whether the voyages of Apollo had actually been completed. This is especially true of the flag-raising at Tranquility Base in 1969, but in other instances light and shadow have been analyzed to raise doubts in the minds of unsuspecting observers. In the end, however, it is the story of American success that is offered through this imagery. It conjures beliefs in the best the nation has to offer.

The spectacular photographs speak to all those who recall Apollo, often with awe and sometimes with the tsk-tsk of what might have been had another course been taken. At no time does the imagery Earthlings have seen stack up to the awesome experience of witnessing it firsthand. At some level, this imagery is a bit like the Albert Bierstadt paintings of the American West in the 19th century, with all their romanticism and idyllic conjurings. On another level, they are akin to the photography of Ansel Adams at Yosemite National Park, with their mystical, intangible, nonmaterialistic quality and their calm and grounded vision.

[7]

Applying Knowledge from
Apollo to This-World Problems

If Americans could be so successful as to land on the Moon, why could
they not solve the other problems of society using the same approach?
Perhaps the lessons of Apollo could be exported elsewhere and used in
solving the myriad public problems taken up by the Johnson administration and its successors. The application of these ideas to city administration, public health, social programs such as Social Security and
Medicare, energy, and veterans' affairs offers a case study in seeking to
control what ultimately proved uncontrollable. It also emphasized the
ability of the federal government to accomplish positive results in social
transformation that have become a part of American culture using space
program management practices. NASA Administrator James E. Webb
incorporated this concept into his 1969 book *Space Age Management:
The Large-Scale Approach*, and acolytes from NASA such as Robert C.
Seamans and Thomas O. Paine sought to apply them elsewhere.

The linkage of space policy and social policy may seem tenuous at
first, but both celebrated the power of the federal government and the
state system to affect in fundamental ways the lives of Americans. During
the 1960s, an activist government emerged to pursue an agenda more
oriented toward change than any other since the 1930s. The era has often
been interpreted as the high-water mark of liberalism, using the federal
government to accomplish far-reaching objectives.

The environment that spawned this sense of change through governmental intervention gained credence with the inauguration of President
John F. Kennedy in January 1961. The administration presented the
image of an energized national movement that sought to use the power
of the federal government to effect change. Presumably, this represented
change for the better, but in the years since the 1960s there has been a

wide divergence of opinion on the outcomes of those efforts. Critiques percolating for many years emerged full-blown during the last two decades of the 20th century to reconsider the history and policy of liberal ideology in the United States. In the process, several scholars and political pundits have reappraised and castigated the social upheaval of the 1960s, defeat in Vietnam, and Great Society programs as failures of American politics. Whether viewed as a positive or a negative, however, Kennedy's administration expressed a strong consensus that science and technology, coupled with proper leadership and the inspiration of a great cause, could solve almost any problem of society.

Examples of governmental activism on the part of the Kennedy administration abound; the war on poverty, the Peace Corps, support for civil rights, the Great Society programs of Lyndon Johnson, and a host of other initiatives are examples. Johnson even tried to defend NASA as a part of his Great Society initiatives, arguing that it helped poor southern communities with an infusion of federal investment in high technology. These all represented a broadening of governmental power for what most in the liberal majority perceived as positive purposes.

Webb, NASA administrator from 1961 to 1968, championed the exporting of the ideas that made Apollo possible to other avenues of action. In the words of political scientist W. Henry Lambright: "While the decision to go to the moon was unfolding, a separate decision process—mainly in Webb's own mind—was unfolding. This was the personal agenda Webb had brought to NASA—a mission to use science and technology . . . to strengthen the United States educationally and economically." Webb sought to create a "Space Age America" that argued for the export of the technocracy and bureaucracy needed for Apollo to address societal needs. He had a broad list of targets: stimulating the economy, advancing education, and applying new management techniques and technologies to solve urban, agricultural, or resource problems. In Project Apollo, Webb saw the seeds of transformation for the nation as a whole. The Moon-landing program combined intensive planning and hierarchical organization with decentralized decision making and innovation. In accomplishing the Moon program, NASA successfully integrated myriad technical and professional cultures with a centralized management structure that applied sophisticated systems management and configuration control.

Perhaps Webb's perspective was to be expected. He entered public service in 1932, serving as a personal assistant to Democratic Congressman Edward W. Pou, from the Fourth North Carolina District, who was also chair of the House Rules Committee. He also later worked for O. Max Gardner, a Washington power broker, attorney, and former governor of North Carolina. President Harry S. Truman asked Webb, after World War II, to serve in his administration. Webb cut his political and philosophical teeth on the pragmatic, innovative, liberal approach to using government power for "public good"—insofar as that could be determined, and those decisions were always controversial—then so much a part of the Democratic Party's raison d'être. His long experience in Washington was very useful during his years at NASA, where he lobbied for the space program and dealt with competing interests on Capitol Hill and in the White House.

By the time of his arrival at NASA in 1961, Webb was a longtime Washington insider who had developed significant skills in bureaucratic politics. Just as he came into the NASA leadership position, Webb wrote to a colleague: "I believe in the Democratic Party as a vehicle for good government and second because I think there is some virtue in consistency. I feel it is better to stick with one organization and try to work through it and within it for the public good." In the end, through a variety of methods, as NASA administrator, Webb built a seamless web of political liaisons that brought continued support for and resources to accomplish the Apollo Moon landing on the "end of the decade" schedule President John F. Kennedy announced in 1961.

Webb's commitment to activism in government for the "benefit of society" found expression in many settings. The most resonant was his campaign to use NASA to catalyze the aforementioned Space Age America. In Webb's conception, the nation would succeed best if society as a whole harnessed science and technology for peaceful and positive purposes, for solving social and environmental problems, and for fostering economic growth and business diversity. At the same time, Webb was one of several engineers, industrialists, social actors, and political leaders in the latter half of the 20th century who embraced technology as the answer to all, or virtually all, of the world's problems.

Webb believed not only that education in the sciences and engineering would be a handmaiden in this effort; they were also, in his estimation, the only salvation for the future. In the end, Webb's Space Age America would be one of unlimited potential. Considerable legislation in the United States is utopian, for it is predicated on the idea that humanity can change the world and make it a better place. For Webb, the space program carried this role forward as well.

He believed fervently that people with vision, skill, and the sense of the greater good, regardless of whether they were situated in governmental or industrial positions, could solve any big problems they might encounter. Remarkably positive, such a philosophy recognized a joining together of the public and private interest and the necessarily associative relationship between them. Webb espoused a general "associative state," similar to Herbert Hoover's vision of public/private partnerships. Webb was, therefore, very much a part of his times. At stake, Webb saw the linkage of Western democratic ideals with social stability, economic and cultural progress, and equanimity. Webb became convinced that only the United States could lead this effort, and to ensure success it required the rise and perfection of a capitalistic economic system that afforded egalitarian opportunity and ensured a robust and growing middle class. The government and the corporate world both had an important stake in this effort, and science and technology were its major vehicles.

Likewise, once at NASA, Webb was persistent in his belief that investment in science and technology would return multifold what was initially spent on them. Less than six months after he arrived at the agency, Webb addressed a memorandum to NASA officials that stated his position on the role of science and tech in advancing the American economy: "One of the most important aspects of the space program is the possibility of the feed-back of valuable, new technological ideas and know-how for use in the American economy." Not long thereafter, he discussed with E. F. Buryan, president of Motec Industries Inc., "the urgent necessity for a strong technological underpinning for any regional economic system that has survival qualities. Indeed, the presence of basic research and the kinds of people who do basic research is of urgent importance for the long run and should be effectively worked out along with the technological and industrial competence." Webb viewed NASA as critical to that economic development. "We are going to spend 30 to 35 billion dollars pushing the most advanced science and technology," he wrote, "endeavoring in every way possible to feedback what we learn into the total national economy."

Even as the While House considered the possibility of undertaking the Moon program in the spring of 1961, Webb was already planning to mobilize resources made available through that effort to tackle the nation's woes. On May 15, less than two weeks before Kennedy's famous speech announcing the decision, he wrote to Vannevar Bush at the Massachusetts Institute of Technology, whom he had known since World War II. Webb knew that Bush was no supporter of human space exploration, and he also knew that getting him to endorse the Moon landings would be a significant boon to the acceptance of the program among those in the scientific community. He told Bush that he regretted that we "find ourselves on somewhat different sides of the complex question of manned space flight." He noted that they had disagreed over the issue at a recent social event and was concerned that the wine-fueled debate had spun out of control. Webb asked him to reconsider his position, adding, "No one could have ridden down Pennsylvania Avenue with Commander Shepard without feeling the deep desire of those lining the Avenue for something to be proud of and a hero. At the moment I believe this feeling is somewhat expanded to include a desire for a real effort in the space field."

This was prelude to his core argument: "In the programs that are now underway and which will shortly be put forward, I expect to do all that I can to build up the university research, teaching, and graduate and post-graduate quality and quantity of education. . . . If we do not find ways to make the major program carry a burden in each of these fields, we simply are not going to meet the challenge of our times." He never won Bush over, but the role of this funding in raising national scientific and technological capabilities, and therefore offering economic advantage for the nation, proved over time to be both true and obvious even to skeptics such as Bush.

Webb never shirked from his belief that economic develop would flow from the Apollo commitment and would therefore remake American society for the better. He argued for this concerning the American Southwest when discussing it with politicians from the region, and he told Vice President Lyndon Johnson that Apollo

would permit us to think of the country as having a complex
in California running from San Francisco down through the
new University of California installation in San Diego, another
center around Chicago with the University of Chicago as a
pivot, a strong Northeastern arrangement with Harvard, M.I.T.,
and like institutions participating. Some work in the southeast
perhaps revolving around the research triangle in North Carolina
(in which Charlie Jonas as the ranking minority on Thomas's
Appropriations Subcommittee would have an interest), and with
the Southwestern complex rounding out the situation.

This approach was a "grand mix of noble vision and pork-barrel politics," as
W. Henry Lambright properly dubbed it, but the value of the Apollo program
to remake society was only heightened through such political emphasis.

Webb continued his crusade on behalf of the role of NASA's science
and technology as a catalyst for economic growth throughout the remainder of his career, badgering governors, mayors, members of Congress,
and business leaders to recognize and pursue, as he put it, "the best ways
of utilizing the tremendous developments of science and technology in
what might be called a total-community-workable-plan kind of concept."
Elsewhere Webb emphasized

that the nation has, through its democratic processes, adopted
an important large program in which many elements of our
society are cooperating, and which is in reality the development
of technology to do exploration and application of new
knowledge. . . . I would like to make the point that we have here
in the national government the means for making decisions
through the representatives of the people and actions by them
based on responsible representations and recommendations by
the President. . . . This program is typical of the rapidly changing
environment of our times, and terminate with the concept that
change is the most important feature of life today.

These were significant attributes of the development and application of
science and technology in modern America and of harnessing them for
the public good through the democratic process.

Webb also recognized that this effort was properly within the province of the federal government as it undertook actions for the public good in the "positive liberal state." In a statement of his position concerning a NASA industrial applications program in 1962, Webb commented:

1. Eighty per cent of the professional and technical personnel engaged in the research and development activity in the United States are working, directly or indirectly, on government programs.
2. Seventy per cent of all research and development expenditures in this country are currently being financed by the government.
   The stimulation of business and industrial growth through application of new knowledge and innovations gleaned from this huge research and development program could be of great significance. It may well assist in achieving the Administration's goal of increasing the Gross National Product at a five per cent rate over the next decade. Moreover, it is incumbent on us, while achieving our specific mission objectives, to make available to citizens generally the specific practical benefits which can flow from a research and development program of this magnitude.

If the macroeconomic studies sponsored by NASA were an indication, as Webb anticipated, the returns on investment in space research and development were astounding. A Midwestern Research Institute (MRI) study of 1971 determined that NASA R&D provided an overall 7:1 return. Essentially, for every dollar spent on R&D, seven dollars were returned to the gross domestic product (GDP). MRI refined its study in 1988, calculating this time an even higher 9:1 return on investment. That was nothing compared to a 1975 Chase Econometrics study, "The Economic Impact of NASA R&D Spending," which reported a whopping 14:1 return on investment.

At the same time, Webb believed that the lessons of management employed at NASA—of course, he viewed himself as central to defining what he believed to be a stellar management structure—would have application to other problems that the nation faced. His approach to "space age management," as he termed it, focused on conducting a symphony of many diverse elements to accomplish truly large-scale and significant objectives. "It seems to me, and I believe you agree," he wrote to nuclear pioneer Lewis Strauss in 1961, "that one of the greatest challenges to democratic government is the ability to carry on large-scale organized effort efficiently. If we don't use our resources efficiently, we simply cannot compete in the kind of whirl we are in." He also laid out this belief for Senator Estes Kefauver of Tennessee in 1963:

As a nation we are in a period of vast and rapid change when we must find better ways to use and guide the giant forces at play to ends that will prevent war and make for a better world. The management systems, the new kinds of relationships we are developing in the government-university-industry field, and the research in methods of organization which we are conducting, have the potential to powerfully reinforce our democratic institutions. I would like very much to endeavor to find, if possible, those constructive areas within which we could work together along this line.

Webb left NASA rather unceremoniously in 1968 in the aftermath of the Apollo fire, while the program was nearing a successful completion. Yet he remained a believer in his program, writing in 1969, "Our society has reached a point where its progress and even its survival increasingly depend upon our ability to organize the complex and to do the unusual." Proper expertise, well-organized and led, and given sufficient resources, could resolve the "many great economic, social, and political problems" that pressed the nation. He viewed the approach to Apollo management as a model of what might be tried elsewhere. Other observers also viewed the management structure of NASA as something special, the central component to success in the Apollo program. Could it be exported to solve other problems in other settings? Webb certainly believed this was possible and tried to replicate its use elsewhere.

One of the earliest efforts of Webb to bridge the divide between NASA's new management approach and other entities working on other problems came with his creation of educational institutions at NASA to enhance the agency's influence in the world beyond. For example, within a few months of the Apollo commitment in 1961, Webb led NASA down one of its most innovative paths of the era. He established the Sustaining University Program (SUP) to employ universities for achieving socio-economic goals. SUP involved a broad-based set of initiatives to increase the size and quality of scientific and technical (S&T) education programs with the goal of reshaping the nation through large-scale S&T activities. Using fellowships, grants, and facilities monies, it aimed at expanding the number of scientists and engineers, especially with advanced degrees, spreading federal largess far beyond previous bounds into geographically and ethnically and racially separated institutions. As Webb explained, he hoped to entice outstanding students to enroll in science and technology programs across the nation rather than at only a few elite schools. He also sought to use these funds for interdisciplinary research projects across a broad spectrum, even including some social scientists in the effort, advancing science and technology and studying its effects. Finally, Webb intended SUP as a strategic, proactive effort to advance society as a whole, with NASA playing a key role.

The Sustaining University Program proved successful in building support for NASA and the Apollo program during the 1960s, in no small measure because large sums of money with relatively few restrictions were made available to colleges and universities nationwide. By 1965, a NASA report noted, "142 universities have received grants to support a total of 3,132 candidates for predoctoral training fellowships in space related fields. Research grants under the Sustaining University Program have been made to 53 educational institutions, most of them involving interdisciplinary effort, and many of them 'seed grants' aimed at strengthening research activity at universities capable of expanding their research programs." By 1970, SUP students had earned more than 4,000 doctorates. Training large numbers of scientists and engineers, which SUP facilitated, epitomized what the program did best.

Additionally, NASA's efforts to reach the Moon helped to inspire a cohort of young people who then went on to become scientists and engineers working in a range of settings that benefited modern society. For instance, *Nature* published the results of a poll in the summer of 2009 that concluded that more than half of the more than 800 scientists polled were inspired to pursue science—and not only astronomy or planetary science—by NASA's Apollo program. "I became completely space crazy," said one life scientist. "I was certain I'd be an astronaut. My interest shifted to biology, but I still believe Apollo 11 was a major influence on me." The study concluded: "More than 80% felt that the life sciences, physical sciences, engineering and human physiology all benefited to some degree from human spaceflight, and almost 90% said that it still inspires younger generations to study science." Others have found similar inspiration through Apollo, and especially how NASA mobilized interest in it to advance science and technology in educational institutions.

While Webb's goal of using universities for the advancement of science and technology succeeded well, his effort to foster NASA-based socioeconomic progress through the same venues proved elusive. As one NASA report concluded in 1968:

> Little evidence was found that the Memorandums of
> Understanding associated with Sustaining University Program
> facilities grants have led to anything but talk. Usually only
> a few administrators with a university even knew about the
> Memorandum. They had not attempted to use it as a tool to induce
> changes in procedures or attitudes; they did not regard it as
> requiring them to do anything new or different.

The same report noted that Webb had intended the SUP as broad-based, involving scientists, engineers, and social scientists. But efforts to persuade university officials "to involve social scientists in their research [was met] with little response. The small amount of social science involvement that does exist is usually on a subproject that does not interact with other research." This involvement, critical to Webb's view of NASA as a transforming force in American society, represented at best a marginal success. Despite Webb's haranguing to anyone in universities who would listen, his crusade to use space technology to raise the spirit and performance of the United States fell far short of his vision.

With the mixed success of Webb's approach to furthering science and technology with NASA support to institutions of higher learning, it may be surprising that he sought to export his ideas to other arenas with as much zeal as he did throughout his career at NASA and even after he left the agency. He immediately latched on to the problems of the city that were apparent as the decade of the 1960s began but became more acute as time passed. In March 1963, he cosponsored the conference "Space,

Science, and Urban Life" with the Ford Foundation, the University of California, and the city of Oakland. It posed two important questions: "(1) Can a national program of space exploration be applicable to the daily tasks of men and women who live and work in our central cities? (2) How may new knowledge, developing in these days of scientific and technological revolution, be used to seek answers to the critical issues of expanding urban populations?"

One reviewer of the published proceedings boiled down the issues discussed at the conference into three broad themes. First, the large-scale effort of NASA to reach the Moon had forced the organization of resources and capabilities on a scale seldom seen in nonmilitary situations, and all participants believed that knowledge gained from that process could be transferred elsewhere from NASA and benefit the nation as a whole.

Second, there was nothing special about technological solutions and their applicability to urban problems. As one participant put it, "Science and technology have done to the city what they have done to any part of human endeavor they have touched. They have freed us more and more from our environment; they have given us more opportunity to manipulate it." Third, the belief prevailed that the federal government would become increasingly involved in the affairs of ordinary citizens, the economy, and the social setting.

These major themes aside, there was little in the proceedings on how science and technology could actually help solve some of the challenges faced by American cities. It might lead, the participants agreed, to cleaner-burning fuels, more efficient automobiles, and better mass transit systems—that is, it might solve problems that are inherently technological. But the truly dicey problems of race, class, economic disparities, and the like defied a technological fix, and almost everyone knew it except for some NASA advocates.

None of this was lost on Webb. He insisted that the nation was in serious trouble, making offhand remarks about urban crises, failing infrastructure, the war on poverty, and a wealth of other social issues. He believed that the only way of meeting these challenges was to organize and coordinate activities on a massive scale. Calm and deliberate responses were necessary, Webb asserted, and he thought that a "multidisciplinary, large-scale effort" must be "more deliberate, more carefully planned, and more interrelated to a multitude of important activities than crisis conditions permit." Drawing lessons from Apollo, the Polaris missile development program, the New Deal's Tennessee Valley Authority, the Marshall Plan for post–World War II reconstruction, and the State Department's postwar reorganization, he argued for a complex management approach to better serve the nation in the face of a complex and changing world. At one level, Webb seemed to advocate, as one reviewer noted,

> for more and better research in the area of management—for better and more dependable ways to organize and administer the great enterprises in which our nation must increasingly engage. This research, according to Mr. Webb, must provide greater knowledge and deeper understanding of what successful leaders in unprecedented endeavors really do, why they do it in a prescribed manner, and what the effects of these actions are.

As an advocate of the use of science and technology in the service of the positive liberal state, Webb proved a persistent, if not entirely successful, apostle. He never got beyond the general ideas of large-scale "space age management" as a panacea for all of the ills of society, but he also never stopped talking about it.

While Webb advocated, some in NASA were uneasy about it. For one, NASA Deputy Administrator Robert C. Seamans Jr. expressed dismay that Webb perhaps "went a little far." Thomas Paine, who succeeded Webb as NASA administrator in 1969, tried to show the commonalities between NASA and the cities as well as their dissimilarities. He noted that science and technology could help deal with many infrastructure issues in cities, ranging from transportation systems to the delivery of city services. But he also found that NASA probably had an easier challenge than did major American cities. NASA had the luxury of defining "specific, stated, measurable goals," but cities' objectives are more diffuse and less measurable. They "have their report card marked against wobbly success standards involving prejudice, special interest, wishful thinking, conflicting values, loose rhetoric prophecy and revelation, or, in the current vernacular—SOUL. A social theory to guide urban society is nonexistent—or worse!" Paine admitted that whatever arguments he might make on the application of space age management to the problems of the cities were abstract but modern management approaches, formalized planning, and strong measures of performance were a step in the right direction.

Indicative of the manner in which NASA's technocratic approach to problem solving interacted with the social issues of the latter 1960s came with the launch of Apollo 11 in July 1969. Rev. Ralph Abernathy, successor to Martin Luther King as head of the Southern Christian Leadership Conference, organized a protest at the launch to call attention to the plight of the poor of the United States. He and 500 marchers of the Poor People's Campaign arrived at the Kennedy Space Center to contest the expenditure of funds for the Moon launch when so many in the United States were ill clothed and poorly housed and fed. The protesters held an all-night vigil as the countdown proceeded, then marched with two mule-drawn wagons as a reminder that while the nation spent significant treasure on the Apollo program, poverty ravished many Americans' lives. As prominent SCLC officer Hosea Williams said at the time, "We do not oppose the Moon shot. Our purpose is to protest America's inability to choose human priorities."

This protest pointed up more effectively than almost anything else the confluence of high-technology challenges and the more mundane but ever-present problems of American society. Abernathy asked to meet with the NASA leadership, and Paine visited the protest site the day before the launch. As Paine recorded the incident:

We were coatless, standing under a cloudy sky, with distant thunder rumbling, and a very light mist of rain occasionally falling. After a good deal of chanting, oratory and lining up, the group marched slowly toward us, singing "We Shall Overcome." In the lead were several mules being led by the Rev. Abernathy, Hosea Williams and other leading members of the Southern Christian Leadership Conference. The leaders came up to us and halted, facing Julian [Scheer, NASA's Public Affairs chief] and myself, while the remainder of the group walked around and surrounded us. . . . One fifth of the population lacks adequate food, clothing, shelter and medical care, [Rev. Abernathy] said. The money for the space program, he stated, should be spent to feed the hungry, clothe the naked, tend the sick, and house the shelterless.

Abernathy said that he had three requests for NASA: that 10 families of his group be invited to view the launch; that NASA "support the movement to combat the nation's poverty, hunger and other social problems"; and that NASA technical people work "to tackle the problem of hunger." Paine acceded to Abernathy's wish and admitted a busload of protestors to view the Apollo 11 launch from the VIP site with other dignitaries. He also commented on how hard it was to apply NASA's scientific and technological knowledge to the problems of society:

> If we could solve the problems of poverty in the United States by
> not pushing the button to launch men to the moon tomorrow,
> then we would not push that button. . . . The great technological
> advances of NASA were child's play compared to the tremendously
> difficult human problems with which he and his people were
> concerned. I said that he should regard the space program,
> however, as an encouraging demonstration of what the American
> people could accomplish when they had vision, leadership and
> adequate resources of competent people and money to overcome
> obstacles. I said I hoped that he would hitch his wagons to our
> rocket, using the space program as a spur to the nation to tackle
> problems boldly in other areas, and using NASA's space successes
> as a yardstick by which progress in other areas should be
> measured. I said that although I could not promise early results, I
> would certainly do everything in my own personal power to help
> him in his fight for better conditions for all Americans, and that
> his request that science and engineering assist in this task was a
> sound one which, in the long run, would indeed help.

Paine then asked Abernathy that when he held a prayer meeting later that day with his protestors they "pray for the safety of our astronauts." As Paine recalled, "He responded with emotion that they would certainly pray for the safety and success of the astronauts, and that as Americans they were as proud of our space achievements as anybody in the country."

Paine concluded that the social problems of the United States could not be solved entirely by revectoring resources from NASA to other initiatives. He also rightly agreed that the problems of society were much more complex and defied resolution using the tools, knowledge, and resources employed to accomplish Project Apollo. While it might be tempting to generalize from the experience of NASA during the 1960s that its success might be duplicated elsewhere, such was not the case. As one observer commented: "NASA's effective implementation of the Apollo mission shows that anything we set our minds to can be done, provided all the conditions are met. Unfortunately, there will be few areas in American life where such will be the case. Nevertheless, Apollo will serve as an everlasting precedent to which optimists will be able to point."

In a manner uniquely ironic, the success of NASA under James Webb showed how malleable and straightforward technological fixes might be accomplished when tackled by skilled leaders with sufficient resources. But they must be questioned whenever they are applied to the task of transforming society. Almost all NASA officials since Webb have agreed that this was true. Space-age management as a concept ended at NASA with the departure of Webb from the scene.

Even so, its legacy has lived on. While much more modest, the application of technological fixes to society's ills using NASA tactics has been successful in some settings. For example, NASA pioneered important efforts to create such things as solar energy, biomedical breakthroughs, and a host of other life-changing technologies arising from efforts in space. NASA calls these spin-offs, commercial products that had at least some of their origins in spaceflight-related research. Most years, the agency publishes a book describing some of the most spectacular, ranging from laser angioplasty to body imaging for medical diagnostics to imaging and data analysis technology. With the caveat that technology transfer is an exceptionally complex subject that is almost impossible to track properly, these various studies show much about the prospect of technological lagniappe from the US effort to get to the Moon and from later space research.

One may point to several significant outgrowths of Apollo that have entered the workplace since the 1960s and are now ubiquitous innovations used everywhere. The most important is the application of the systems management concept to practical problems in normal life. Pioneered by the US Air Force in the 1950s, especially in relation to the development of ICBMs, NASA developed and popularized centralized authority over design, engineering, procurement, testing, construction, manufacturing, spare parts, logistics, training, and operations. A fundamental tenet of the program management concept was that three critical factors—cost, schedule, and reliability—were interrelated and had to be managed as a group. Many also recognized these factors' constancy; if program managers held cost to a specific level, one of the other two factors—or both of them to a somewhat lesser degree—would be adversely affected. This held true for the Apollo program and for any other development effort. Accordingly, lessons from Apollo went into complex management projects thereafter. The use of sophisticated diagrams and tracking mechanisms, the fitting out of a control room that looked remarkably like NASA's Mission Control Center, the organization of personnel, resources, time, and circumstances as thoroughly as possible all found their way into civil society from the space program. Most important, the organization and communication of knowledge, as done in Apollo, was replicated in other major endeavors. The result is an approach to disaster preparedness and management, to urban planning, and to other civil projects that looks surprisingly like Apollo and its conduct by NASA engineers in the 1960s.

Even so, what was called space-age management as a concept was a product largely of the mind of James Webb, and it was advanced through the force of his indomitable will. Little of substance came from it, and even less persisted beyond the Apollo era. In its time and place, however, it found some adherents, although most were modest in their claims for it. Technological fixes found expression in certain decidedly technological concerns in America's urban areas, but certainly not elsewhere. In the end, most would probably agree with the assessment of William Hines in a column in the *Birmingham News*: "[NASA] could be depended on to give a good account of itself in purely scientific or engineering situations, [but] its ability to handle problems with a big 'people' component is largely untested." He added that Wernher von Braun remarked at a social setting "that getting to the moon was easier than getting to his table in the ballroom because down here on Earth there are always people in the way."

There do seem to be some good things that the positive liberal state has accomplished. There are others that have thus far eluded its best practitioners. One might draw several conclusions from this discussion. One is that there is no hope of large-scale management of social problems that has much chance of success. One would do well to avoid leaping to such a determination. Another might be that space-age management as a concept advocated by James Webb was flawed. Such might be the case, especially when seeking its universal rather than its limited application. Others might question the nature of the positive liberal state as a whole, but this condemns Americans to a society without promise and hope for a better place. Instead, Apollo's success offers the lesson, sometimes well learned but often not so much, that well-defined, well-ordered, and well-orchestrated projects can be accomplished when sufficient leadership and funding is available.

In an era strikingly different from the 1960s, the example of Apollo still resonates and serves as an example for what might be accomplished with well-supported actions taken through inspired leadership. As recently as 2009, the time of the *Nature* poll, certain core beliefs remained about the Apollo program and the export of its ideas to other problems, in this case the challenge of global climate change:

> The Apollo Analogy tells us that we (the United States and, well, per-haps humanity) can organize to achieve something greater than ourselves.
> The Apollo Analogy tells us that we can surprise ourselves by the rapidity and extent of change, if we set ourselves to it.
> The Apollo Analogy tells us that stretch targets, objectives that seem next to impossible, can lead us to do better than imagined.

A related comment set the tone for this recollection of Apollo: "There is an intense power to the appeal to the greatness of the Apollo Program's quite tangible achievements as a model for tackling the very serious challenges before us (both the U.S. and all of us). . . . The Apollo Analogy tells of the power of vision . . . and calls to meet it."

Likewise, on January 24, 2011, Senator John Kerry of Massachusetts invoked the Moon landings in calling for efforts to deal with a lethargic economy when he asked that America seek a new Apollo program, one by which we would "go to the moon here on Earth in our push for technology," adding, "We're going to make the same kind of commitment to be first, to lead the way, and that is the single biggest kick I can think of to the American economy."

An irony of the first magnitude cannot be ignored here. Efforts to apply knowledge gained in the accomplishment of the Apollo program during the James Webb era to other problems of society had mixed results at best. Yet Apollo as symbol has remained always as a hopeful suggestion that solutions may be found. NASA's engineers and scientists went to the Moon, one of the hardest tasks ever accomplished; might we be able to solve equally challenging problems in the future in a similar manner? Dealing with global climate change, the search for new sources of creating and delivering energy, an exponentially expanding world population, an emerging scarcity of water, the problems of economic inequity and competitiveness, all have some potential for the application of technical solutions. But they are largely political and social problems. Will they also be solvable using those lessons from Apollo? In other words, "If we can put a man on the Moon, why can't we do X?"

8

Apollo and the Religion of Spaceflight

What if we viewed the history of Apollo somewhat less through the lens of Cold War politics and more as an expression of what might be called a religion of spaceflight? There is a deeply religious quality to advocacy for the investment in and support for human space exploration, lending to the endeavor a "higher purpose" that helps to explain both the generous nature of the actual investment and the ultimate unwillingness of Americans to eviscerate space budgets despite less than full support for extraterrestrial travel.

The word *religion* carries several connotations, chief among them being the practice of faith and worship, the existence of a set of beliefs inspiring reverence and allegiance, trust in an alternative arrangement of human affairs that cannot be physically demonstrated, a frequent promise of immortality, an explanation of origins, and conviction in a message of salvation. Human space exploration fits these characteristics well and represents what might be referred to as a form of civil religion in the United States. Although the term extends back to Jean-Jacques Rousseau, sociologist Robert Neelly Bellah argued in 1967 that Americans possess a keenly developed civil religion that accepts a common set of values, principles, rituals, memories, holidays, and beliefs. These reside parallel to more traditional religious ideals and inspire faith, worship, and reverence and offer salvation. He asserted that there was "an institutionalized collection of sacred beliefs about the American nation" that involved "national self-worship" and called for "the subordination of the nation to ethical principles that transcend it in terms of which it should be judged." Such a belief system addresses deep-seated needs that strike at the confluence of the scientific pursuit of knowledge and the philosophical understanding of humanity's place in the universe.

Previous observers have hinted at the embrace of flight by a segment of humanity as essentially a form of religion. The ability to fly represented the opportunity to transcend the earthly realm and reach a "higher plain," something that many viewed as both a romantic and a religious experience. Religious studies scholar Charles Reagan Wilson suggests that Apollo participants tied religion and science together in a surprising way, concluding, "One sees elements of both a rational religion, which can be traced back to the Enlightenment, and an evangelical religion, which has been the nation's culturally dominant religious force through most of American history."

Like a religion, space exploration receives special treatment that is accorded to only a few American movements. Perhaps this fact helps to explain why Apollo occupies such a central place in American culture and receives the level of public support that it does.

Without question, those associated with spaceflight have spoken of it in explicitly religious terms. For example, Chris Kraft, a leading NASA official during the Apollo era and director of the Johnson Space Center in the 1970s, characterized his support of space exploration in overtly religious language: "This step into the universe is a religion, and I'm a member of it." Others have done so as well, sometimes garnering negative reactions. As JFK staffer Charles Johnson wrote to National Security Advisor McGeorge Bundy on May 21, 1963, "There is already too much religion in the space program." Such expressions were not limited to the United States. In a report about the Soviet space program, H. A. van Helb, the Dutch ambassador in Moscow, commented about of spaceflight in the USSR: "It is significant that a regime which preaches atheism above all else, cannot do without heaven, in a way."

From the beginning, Apollo represented as much as anything else a spiritual quest, a purification of humanity, and a search for absolution and immortality. At a uniquely oblique angle, Apollo was the incarnation of a new religious conception, and in fact the connections between "conventional" religions and spaceflight are very real. The fact that spaceflight might be itself a new religion is especially telling. Norman Mailer eloquently expressed the linkage of these elements at the time of the Apollo program, saying the missions were really about seeking to become one with God:

They don't know what to do when they get there. The fact
that it's technological is what's wrong with it. It's too
exclusively technological. People are sick to death of technology.
The technologists themselves are wondering how they can control
technology before technology wipes out the Earth. So what we're
looking for at this time in human history is an enlargement of
human consciousness, a rediscovery of spiritual values to which
we can adhere because they deepen us.

Apollo evoked, in a metaphorical and absolutist sense, emotions
of awe, devotion, omnipotence, and, most important, redemption for
humanity. It embodied a new clerical caste (the engineers and especially
the astronauts); arcane rituals (the doings of Mission Control and other
operational activities) that were deeply mythical as well as imbued with
higher purpose; a language of devotion (the NASA jargon invoked by
both practitioners and acolytes/enthusiasts); articles of faith (expressed
in such exemplars as the "destiny" of humans to colonize Mars); and a
theology of salvation that allowed humanity to reach beyond Earth and
settle the cosmos. The promise of a utopian Zion on a new world, coupled
with immortality for the species, resonates through every fiber of the
space exploration community. Wernher von Braun, as only one example
among many, viewed space exploration as a millenarian new beginning
for humankind. Such deep-seated convictions energized space exploration and the subjugation of the universe even before the dawn of the
space age.

The prospect of becoming immortal, at least as a species, presumably made possible by Apollo, is compelling. Survival of the species is
the only rationale for spaceflight that ultimately has held sway among
all the others that have been invoked. In essence, it is a salvation doctrine,
one that says that if humanity does not become multiplanetary, it cannot
survive. Carl Sagan wrote movingly about the "last perfect day" on Earth
before the Sun fundamentally changes and ends our ability to survive on
this planet. In their 2002 astrobiology book *The Life and Death of Planet
Earth*, scientists Peter Ward and Donald Brownlee describe the natural
life cycle of stars such as our Sun and the planets that circle them, sketch-
ing out several possible scenarios for the end of life on Earth. Yet they
insist that life on this planet will definitely end when the Sun, having

used up too much of its hydrogen, becomes a red giant and heats Earth until every living thing, no matter how deep underground, is dead.

While this scenario will indeed happen—billions of years in the future—any number of catastrophes could end life on Earth before that time. A much earlier and quite likely way for life, or at least life as we know it, to end is the way it almost ended 65 million years ago, when an asteroid or a comet crashed into Earth. The effects of this collision may have caused the extinction of the dinosaurs and probably two-thirds of all other life on Earth at that time. Yet enough life survived the harsh environmental aftermath to give rise to mammals, a highly adaptable species that even survived the last Ice Age. Throughout history, other asteroids and comets have struck Earth, and a great galactic asteroid between six and nine miles wide left a crater 186 miles wide in Mexico's Yucatán Peninsula, probably killing the dinosaurs. This reality entered most people's consciousness in July 1994, when humans for the first time witnessed the devastating impact of Comet Shoemaker-Levy 9, which crashed into Jupiter with spectacular results.

In time, another comet or meteoroid will again hit Earth with disastrous consequences. Efforts to catalogue all Earth-crossing asteroids, track their trajectories, and develop countermeasures to destroy or deflect objects on a collision course with Earth are important, but to ensure the survival of the species, humanity must build outposts elsewhere. So say those expressing the salvation ideology of the religion of spaceflight. Astronaut John Young once paraphrased the *Pogo* comic strip: "I have met an endangered species, and it is us." None other than Wernher von Braun emphasized the salvation narrative in a 1976 speech to the National Space Institute, pointing to a bright future for humanity if it embarked on the high frontier of space, and death if it did not seize the opportunity. Space, he said, would "offer new places to live—a chance to organize a new interplanetary society, and make fresh beginnings."

Does this belief in survival of the species through emigration to other bodies in the solar system represent a salvation theology? This is a question worth debating, but it is so pervasive among adherents and so persistently advocated by them that it seems to be a near-universally held belief. Furthermore, like those espousing the immortality of the human soul among the world's great religions—their fundamental salvation theology—such statements of humanity's salvation through spaceflight

are fundamentally statements of faith predicated on no knowledge whatsoever. We have no idea whether humans can avoid extinction in this
manner. Indeed, the movement of fragile *Homo sapiens* into a realm
for which they are ill adapted requires the building of artificial systems
that enable them both to survive and to complete useful work. As space
scientist Vadim Rygalov of the University of North Dakota remarked,
"Spaceflight is first and foremost about providing the basics of human
physiological needs in an environment in which they do not exist." From
the most critical—meaning that its absence would cause immediate
death—to the least critical, these requirements include such constants
available on Earth as atmospheric pressure, breathable oxygen, suitable
temperature, drinking water, food, gravitational pull on physical systems,
radiation mitigation, and others of a less immediate nature.

The design of every spaceflight vehicle, every spacesuit, every subsystem of even the most simple design takes as its raison d'être the
task of protecting the astronauts because of the extreme hostility of
the space environment. Absent the discovery of an Earthlike planet to
which humanity might migrate, this salvation ideology seems problem-
atic, a statement of faith rather than knowledge or reason. It remains,
however, a powerful aspect of the religion of human spaceflight.

Apollo also meets another criterion of religion: it provided both saints
and martyrs. The bravery of the astronauts touched emotions deeply
seated in the nation's experience of the 20th century. Apollo astronauts
were seen as humanity's exemplars in the pursuit of the quest of becoming a multiplanetary species. In the United States, they were imbued with
this responsibility in 1959 and have carried it into the present. No example
is more poignant than those lost in the Apollo fire, who are held in reverence and whose memorialization has known few bounds. Virtually every
exhibit concerning human spaceflight commemorates this crew lost in
the exploration of space, and all are deeply reverent, celebrating the bravery of the crew and enshrining the memory of sacrifice. In the National
Air and Space Museum in Washington, DC, there are two places where
those lost in the endeavor are remembered. The first is in the "Apollo
to the Moon" exhibition, in which the Apollo crew lost in a capsule fire
in January 1967 and two Soyuz crews in 1967 and 1971 are pronounced
heroes. The second is in the Boeing Milestones of Flight Hall, where the
exhibition on Apollo invokes them as martyrs to a noble cause.

Despite its inglorious ending in 1972, Apollo remains one of the most popular subjects displayed at the Smithsonian Institution National Air and Space Museum, which often receives more than 8 million visitors per year. Here a crowd gathers in the Milestones of Flight gallery to hear about the space race from docent Bobbie Dyke. (Photograph by Eric Long, NASM image no. 99-15233.3)

Much like the confession of sin in traditional religious traditions, a desire to partake in the guilt associated with the loss of astronauts is a significant part of the religion of spaceflight. For the devoutly religious, the mantra is that humanity is sinful and every individual must accept this fundamental truth. The unworthiness everyone must inculcate into the very depths of their soul can only find release in the salvation achieved through oneness with God. In the Christian tradition, we are ultimately responsible, every one of us, for the death of Jesus Christ. We are worthy only of the pit of hell, but God grants salvation not because we deserve it but because of mercy. Only through human acceptance of this reality, repentance for the evil that inhabits us, and mercy from the throne of God may salvation be attained.

Former NASA Flight Controller Wayne Hale's eloquent statement of his own culpability in the *Columbia* shuttle disaster comes to mind: "I cannot speak for others but let me set my record straight: I am at fault. If you need a scapegoat, start with me. I had the opportunity and the information and I failed to make use of it. We could discuss the particulars: inattention, incompetence, distraction, lack of conviction, lack of understanding, a lack of backbone, laziness. The bottom line is that I failed to understand what I was being told; I failed to stand up and be counted." Such acceptance of responsibility for the death of astronauts was nothing less than the confession of Saint Augustine for his sin and the seeking of forgiveness and salvation.

Equally significant is the heroism of astronauts lost in space exploration. Film director James Cameron has commented eloquently about both the details of accidents in NASA history as well as the larger issue of balancing "the yin and yang of caution and boldness, risk aversion and risk taking, fear and fearlessness. No great accomplishment takes place, whether it be a movie or a deep ocean expedition or a space mission, without a kind of dynamic equipoise between the two. Luck is not a factor. Hope is not a strategy. Fear is not an option."

Alternatively, there are those condemned by adherents of human spaceflight. In the same way that Christians have singled out perceived heretics, those embracing the ideology of space exploration have consigned some to outer darkness. Take the instance of President Richard M. Nixon. In 1970, he torpedoed NASA's planned approach to long-term space exploration after Apollo. Advocates of an aggressive space program have accordingly consigned Nixon, more than any other political leader, to a special place in hell. They did not do so because of his violation of the Constitution in the Watergate affair or because of his lying about sending US troops into Cambodia or because of his destabilization of regimes worldwide if they failed to toe the American line. They did so because he refused to endorse the Space Task Group report when it might have led to an expansive vision for human spaceflight through the remainder of the 20th century.

Others have suffered similar fates, though none has received such widespread denunciation from the space community. One example of such a lesser villain is Rocco A. Petrone, who served as the third director of NASA's Marshall Space Flight Center from January 1973 to March 1974. He presided over a post-Apollo downsizing at the center, officially characterized as "a major center restructuring to accommodate Marshall's changing roles and responsibilities in the 1970s." Less politely, he was condemned as the "hammer" sent to Marshall to uproot the German rocket team that had dominated the center during the leadership of Wernher von Braun between 1960 and 1970. Never mind that such a takedown was never his core objective. One longtime MSFC employee told me that for many years he had waited for Petrone's death, which finally came in 2006, so that he could "piss on his grave," but then he realized that standing in the line for that last act of defiance would require too long a wait.

To suggest that there is such a thing as sacred texts or holy writ in this civil religion of spaceflight may seem incongruous, but I believe such are present and are everywhere invoked by adherents throughout America. One such set of writings that are imbued with this sacredness are the *Collier's* series of articles that appeared between 1952 and 1954, each expertly illustrated with striking images of human spaceflight by some of the best artists of the era. The first issue of *Collier's* devoted to space appeared on March 22, 1952, a headline asking readers, "What Are We Waiting For?" The editors urged Americans to support an aggressive space program. They suggested that space flight was possible, not just science fiction, and that it was inevitable that humanity would venture outward. It framed the exploration of space in the context of the Cold War rivalry with the Soviet Union and concluded, "*Collier's* believes that the time has come for Washington to give priority of attention to the matter of space superiority. The rearmament gap between the East and West has been steadily closing. And nothing, in our opinion, should be left undone that might guarantee the peace of the world. It's as simple as that."

Wernher von Braun led off the *Collier's* issue with an impressionistic article describing the overall features of an aggressive spaceflight program. He advocated sending an artificial satellite into orbit to learn more about spaceflight, followed by the first orbital flights by humans, development of a reusable spacecraft for travel to and from Earth orbit, building a permanently inhabited space station, and finally human exploration of the Moon and planets by means of spacecraft launched from the space station. Several other writers then followed with elaborations on various aspects of spaceflight, ranging from technological viability to space law to biomedicine. The series concluded with a special issue of the magazine devoted to Mars, in which Von Braun and others described how to get there and predicted what might be found based on recent scientific data.

The *Collier's* series of articles in the 1950s represent one example of texts taking on the semblance of holy writ. Over time, such texts are invoked, glossed and reglossed by adherents, and debated as to nuances of meaning. Their influence is unmistakable. Every history mentions them, quotes them, and favorably comments on them. They have been reprinted, used in speeches, and invoked in public policy debates. Finally, they were central to the development of NASA's plans, and have continued to affect strategic thinking about space exploration, especially as it relates to a systematic, stepwise methodology for exploring the solar system.

Spaceflight also has many rituals, and like all rituals everywhere, they have gained a special place in the lives of adherents. They are often complex, sometimes esoteric, and always comforting those who follow them. Like rituals in other religious settings, they are loaded with meaning and symbolism. Key among them is the experience of the launch; the process whereby the astronauts prepare for their missions is highly ritualized. It consists of arrival at the Kennedy Space Center in Florida; isolation and preparation; a ritualistic breakfast; suiting up in a special facility complete with oversized stuffed chairs for the astronauts to lounge in; the walk out to the transport vehicle that carries them to the launch site; riding the elevator to the top of the gantry; and then entering the spacecraft from the White Room, where the astronauts are strapped in and handshakes are exchanged all around.

No work captures the ceremonial connotations of spaceflight more effectively than the 1995 feature film *Apollo 13*. Directed by Ron Howard from a screenplay by William Broyles Jr. and Al Reinert, it tells the harrowing story of Jim Lovell and his crew of Apollo astronauts, who failed to reach the Moon but did return safely to Earth despite a major accident that crippled their spacecraft. Howard produced, whether intentionally or not, a fable that depicted well the very real ritual of preparing for launch. In the film's launch sequence, Howard depicts the astronauts as reverent supplicants seeking higher purpose. Their suiting up for the mission, done with inspiring full-orchestral accompaniment, bears close resemblance to priests donning their vestments in anticipation of spiritual renewal. The spiritual context for this mission continues as Apollo 13 is launched, again accompanied by rousing and heroic music.

This film, as well as the actual experience of launch, was an epiphany for astronauts, launch controllers, and spectators. Some watched in awe; others sobbed with emotion; all were moved by the experience. It represented a scene of redemption for all nonbelievers. At the conclusion of the launch sequence, with Apollo 13 safely in orbit around Earth, Mission Control comments, like the priest at the conclusion of a Mass, "And that is how we do that." At a fundamental level, the launch sequence is a human communion with deity. To release such energy under total human control is to become like the gods, to transcend the earthly plane and reach for heaven.

Likewise, the actual experience of any launch has long been viewed in deeply religious terms. It represents a pilgrimage for all who participate, traveling long distances and enduring the vicissitudes of weather, malfunction, or other difficulties in waiting for the launch. Watching a launch brings with it the joy of release at the moment the rocket begins its flight, and it sends back viewers changed by the experience. Novelist Ray Bradbury eloquently captures this epiphany in a tone reminiscent of a jeremiad:

> Too many of us have lost the passion and emotion of the
> remarkable things we've done in space. Let us not tear up the

future, but rather again heed the creative metaphors that render space travel a religious experience. When the blast of a rocket launch slams you against the wall and all the rust is shaken off your body, you will hear the great shout of the universe and the joyful crying of people who have been changed by what they've seen.

Bradbury firmly believed that no one leaves a space launch untransformed. Like the Eucharist, the ritual of the launch offers a recommitment to the endeavor and a symbolic cleansing of the communicant's soul. The experience, as he commented repeatedly, is both thrilling and sanctifying. Legendary journalist Walter Cronkite also captured this sense of spaceflight as spiritual renewal in a remarkable reflection written at the turn of the 21st century. "Yes, indeed, we are the lucky generation," he wrote. In this era, we "first broke our earthly bonds and ventured into space. From our descendants' perches on other planets or distant space cities, they will look back at our achievement with wonder at our courage and audacity and with appreciation at our accomplishments, which assured the future in which they live."

There seems also to be a strong group identity among advocates for an aggressive program of human space exploration, akin to the sense of belonging to a religious group. To be a member of this community is to be accepted and trusted, at least until that trust is betrayed, and once in it represents essentially a lifetime commitment. The social networks of this community have long been interlocking and self-reinforcing. The professional linkages are in the same category. And this has been the case since the beginning of the space age.

As only one example, millions of people every year visit locations involved in, or celebrating, spaceflight. The National Air and Space Museum of the Smithsonian Institution in Washington, DC, is the world's most visited museum, hosting more than 8 million tourists each year. An additional 2 million people annually visit the Kennedy Space Center in Florida, making it one of the largest attractions in a state known for its tourism industry. Each year another 2 million tourists visit Space World, a space-oriented theme park, in Kitakyushu City, Japan. Other terrestrial locations, including the various Space Camps, account for a further 2 million visitors annually. Human spaceflight adherents interact with each other in these and other settings and reinforce their beliefs and commitments. These shrines of spaceflight provide the context for connecting one to another in this community.

Collectively, the spirit of Apollo as religion created and nurtured in these settings emphasizes a belief in "chosenness." This is among the most powerful and persistent conceptions motivating enthusiasts for the religion of spaceflight. Adherents have special understanding and it manifests itself through this faithfulness in the possibilities of human space exploration. It has its roots in Old Testament concepts and fosters the conclusion that this endeavor and those engaged in it have a special challenge and opportunity. There is nothing new about this. The author of Deuteronomy recorded that God told the ancient Jewish people: "The Lord your God has chosen you out of all the people on the face of the earth to be his people, his treasured possession" (Deut. 7:6, NIV).

Spaceflight adherents have appropriated this belief as a chosen people venturing into the unknown in a form of "manifest destiny." Historian Walter A. McDougall concluded that Old Testament traditions "were coherent, mutually supportive, and reflective of our original image of America as a Promised Land." Over time spaceflight adherents rendered absolute the idea of chosenness. This was really a part of a larger theme in American history. Neo-orthodox theologian H. Reinhold Niebuhr concluded in 1937: "The old idea of American Christians as a chosen people who had been called to a special task was turned into the notion of a chosen nation especially favored. . . . As the nineteenth century went on, the note of divine favoritism was increasingly sounded." So, too, with the spaceflight community.

So what does all this mean? After a more than half a century's gestation, it is now apparent that space is central to the culture of the United States, and Apollo is even more central, but the reasons for the centrality may have more to do with subtleties and the sublime than with anything else. That may well have been true during the space race, but it has become abundantly clear in the post–Cold War era.

9

Abandoned in Place

All that is left of the Kennedy Space Center's Launch Complex 34 (LC-34)—the first site used to launch the Saturn 1B rockets for the Apollo program in the 1960s and the setting of NASA's first fatal accident, in 1967 during a ground test of Apollo 1—is a concrete and refractory brick pad, a reinforced concrete pedestal that served as a base for the rockets, a launch control center now used for storage, and a steel blast deflector. NASA used the site from 1961 to 1968 and finally mothballed it in November 1971, cannibalizing most of its useful components in April 1972. Even so, NASA retained control of LC-34, and it became a tour stop for visitors to NASA's Kennedy Space Center. Most significantly, LC-34 was designated a National Historic Landmark, along with several other sites at Cape Canaveral Air Force Station, Florida, in April 1984. It now shows significant signs of deterioration from the weather, and no one seems to be actively preserving the site.

When I first visited LC-34 in 1992, I was struck by the desolation of the site. There are three markers, all of them weathered and showing signs of neglect, each suggesting the myth and memory of Apollo. Two of them are plaques documenting the tragedy of Apollo 1, the capsule fire that killed astronauts Gus Grissom, Ed White, and Roger Chaffee, on January 27, 1967. One reads: "They gave their lives in service to their country in the ongoing exploration of humankind's final frontier. Remember them not for how they died but for those ideals for which they lived." The second states, "Ad Astra Per Aspera [A rough road leads to the stars]" and ends "Godspeed to the crew of Apollo 1." Both plaques commemorate, in an especially reverential manner, the most dramatic event of the launch pad's history and the most tragic incident of the entire Moon-landing effort, interpreting the event as an example of sacrifice for a higher purpose and stating faith in the eternal place of the crew in human history. They also suggest that the long road of exploration requires the commitment of those engaged in it, some offering the

As NASA began to launch humans into space, the Kennedy Space Center began to emerge as a tourist destination where people could view the launch complex and learn more about its activities. NASA still operates the visitors' complex as a means of outreach. This brochure shows the center during the Apollo era. (NASA image, Kennedy Space Center Historical Archives)

ultimate sacrifice, and only through this process may America reach for the stars.

The third sign, which is not an official one, offers an epitaph on the whole of the effort. Stenciled on one leg of the concrete pedestal facing the Atlantic Ocean is a simple red declaration: "Abandon in Place." In such a statement, an unknown worker encapsulated the fate of one of the largest and most extraordinary endeavors in the history of the United States, indeed in the history of the world. Ironically, for all of the effort that went into the Apollo program, upon its successful completion much of the infrastructure created to support it was abandoned, some was altered for other uses, and much more was torn down. This includes not only sites on Earth but also six landing sites on the Moon.

What is the place of the cultural relics of Apollo in modern American society? There are, of course, well-cared-for museums and historic sites around the country directed toward telling the story of spaceflight. Apollo always looms large in their exhibits. But which story do they tell? Why do they tell it the way that they do? What do their visitors take away from the experience? Equally important are the relics that have not received the care they might deserve. The rusted missiles and spacecraft placed in rocket gardens around the nation; the Apollo boilerplate capsule in a park gazebo near Lancaster, California, forgotten by most; and the oddities collected by roadside sellers and the ad hoc "museums" with the stray Apollo object are also part of this story. So, too, is the "Moon Hut" in Cape Canaveral, a legendary astronaut hangout since Apollo, and the McDonald's near the Johnson Space Center in Houston, with its giant plastic Apollo astronaut standing on its roof.

Because of both the anomalous nature of Apollo and the technical culture of NASA, which prized everything new and nothing old, there was little value placed on maintaining in working order the infrastructure that made the Moon landings possible. NASA therefore sought either to dispose of these relics with all dispatch or to alter them for other uses. The Apollo Launch Complexes—LC-39A and B—located at the Kennedy Space Center near Titusville, Florida, are the first case. During an aggressive construction effort in the mid-1960s, NASA had built these massive and complex structures as the launch sites for the mighty Saturn V Moon rockets. These launch complexes were designed to support mobile launch operations, with the Apollo/Saturn stack assembled and checked out in the protected environment of the Vertical Assembly Building (VAB), then transported by a massive tracked vehicle to the pad for final processing and launch. Initially, NASA had envisioned five launch complexes along the coast to the east of the VAB that were evenly spaced 8,700 feet apart to avoid damage in the event of a pad explosion. Two of these were to have been developed slowly over time, as the space launch market built up, with the first three to go into service for the initial Apollo program. NASA finished only two of them, numbering the pads LC-39A and LC-39B.

LC-39A and B each had a Launch Umbilical Tower (LUT) that provided access to each part of the Saturn V, from the S-IC stage at the bottom of the stack to the Launch Escape System at the top. Nine service arms extended to the umbilicals of each stage, supplying the stages with fuel, electricity, air-conditioning, and all other consumables, as well as providing 18 access points to the rocket and spacecraft on several levels.

Standing more than 380 feet to the top of the tower (398 feet, 9 inches to the top of the crane), the LUT's most famous feature was the White Room, where astronauts entered the Apollo command module. It was in this famous room that Guenter Wendt presided over every human space-flight mission of the Apollo era. Astronauts called him "Pad Führer," not always with affection. Wendt's emphasis on successfully completing the mission, ensuring the safety of the astronauts, and creatively sidestepping bureaucracy did earn him the admiration of many, though. Many astronauts recall how Wendt strapped them into their capsules, shook their hands, offered words of support, and closed the hatch, the last person they would see before their trip into space. In those moments, they were thankful for his abrasive attention to detail and his forceful leadership on the launch pad.

Launch Complex 39A was decommissioned in 1974 and reconfigured for Space Shuttle operations thereafter. LC-39B underwent a similar Apollo-Saturn decommission in 1977, but it did not enter Space Shuttle operations until 1986, when its first launch was the ill-fated STS 51-L mission, the *Challenger* accident. In the process, some of the older Saturn V superstructure was discarded. NASA stored it in a vacant lot at the Kennedy Space Center, where it rusted for years. NASA workers did remove the most historic part of the LUT, the White Room, and preserved LC-39A's at the Kennedy Space Center Visitor's Complex and the other at the Kansas Cosmosphere in Hutchinson. Even as this took place, LC-39 received designation as a historic site, later receiving a listing on the National Register of Historic Places.

But what of the LUT's remains? Periodically, requests to save the LUTs have emerged, only to collapse when funding, lack of interest, inertia, government stonewalling, or whichever other reason one might care to name overcomes the impulse. NASA has sought to destroy and dispose of the LUT since the early 1980s. Preservationists came forward to try to save LC-39B when it was to be converted for shuttle operations, asking that it be maintained with a Saturn V stack on it as a historic site. Enlisting the support of then Rep. Don Nelson, a Democrat from Brevard County, Florida, where the Kennedy Space Center is located, a grassroots campaign in 1983 to save the complex failed to raise the funds necessary to dismantle and move the LUT to another site. Three years later, this same local effort, by then incorporated as the not-for-profit Apollo Society, tried to raise money to move and erect the LUT at the KSC Visitors' Complex. Its 500 members raised about $50,000 for this purpose, but that was far short of the more than $15 million estimate for the creation of such a historic site.

In early 2004, the latest effort to save the LUT got underway. NASA was preparing to scrap its remains after the Environmental Protection Agency had discovered its orange gantry paint at the open-air storage site was leaching heavy metals and toxic substances into the local water table. NASA decided to spend $2 million to decontaminate and dispose of the pieces, but again preservationists stepped forward to prevent it. Burton Summerfield, chief of the safety, health, and environmental division at KSC, sounded the position of most in NASA when he talked sympathetically of these efforts. "A lot of people have tried over the years to save the tower, but unfortunately no one has come up with the financial wherewithal to do it. For us right now, this is an environmental issue rather than a historic preservation issue," he commented. A hastily organized Space Restoration Society, led by enthusiastic but cash-poor advocates, appealed to save the LUT as something akin "to the piers from which Christopher Columbus set sail in Palos de la Frontera, Spain, with the *Nina*, *Pinta* and *Santa Maria*." Of course, those piers were not preserved either. This preservation effort failed as well. On February 12, 2004, NASA began disposal work on the LUT, and within a few months it was gone.

The story of the Vertical Assembly Building (VAB) at the Kennedy Space Center is quite different. Built beginning in 1962 as the location for stacking the Saturn V spacecraft in preparation for launch, it was as tall as a 40-story building and had 50 percent more volume than the Pentagon. As it turned out, the VAB incorporated several major features that proved useful for launch operations beyond the Apollo program: back-to-back high bays allowing Space Shuttle vehicle erection and assembly without any restrictions, three large cranes for moving components, and capabilities for handling of boosters and upper stages. The VAB served well throughout Apollo, and when the time came to transition to the Space Shuttle program, NASA simply changed the name of the building to the Vehicle Assembly Building while retaining the acronym and made a few changes to the inside of the building for the new space vehicle. At no point has NASA attempted to preserve the VAB as a historic site. It is a working location that looks from the outside much like it did when first erected in the 1960s, dominating the landscape in much the same way it did during Apollo, although it did require some considerable repair following damage resulting from Hurricane Frances in September 2004. Given the retirement of the Space Shuttle program in 2011, one of the high bays of the VAB was reconditioned for use by the SpaceX Corporation in preparation for launches of its Falcon 9/Dragon capsule launch combination, and another will be used by the next major launcher, the Space Launch System (SLS), when it flies beginning in the 2020s.

Yet another approach to the preservation of historic relics of Apollo was taken with the Mission Control Center (MCC) at Johnson Space Center in Houston. First used in 1965 during the Gemini IV flight, when Ed White undertook the first American spacewalk, the MCC served as the control site for all human spaceflight missions through 1998. It was the home of all Gemini, Apollo, Skylab, ASTP, and Space Shuttle missions until that time, and has enormous significance as a historic site. As the National Park Service stated in its "man in space" theme study in 1984:

The Apollo Mission Control Center is significant because of its close association with the manned spacecraft program of the United States. This facility was used to monitor nine Gemini and all Apollo flights including the flight of Apollo 11 that first landed men on the moon. After the end of the Apollo Program this facility was used to monitor manned spaceflights for Skylab, Apollo-Soyuz, and all recent Space Shuttle flights.

The support provided by the Apollo Mission Control Center to the first manned landing on the surface of the moon was critical to the success of the mission. It exercised full mission control of the flight of Apollo 11 from the time of liftoff from Launch Complex 39 at the Kennedy Space Center to the time of splashdown in the Pacific. The technical management of all areas of vehicle systems of Apollo 11 including flight dynamics, life systems, flight crew activities, recovery support, and ground operations were handled here.

Through the use of television and the print news media the scene of activity at the Apollo Mission Control during the first manned landing on the moon was made familiar to millions of Americans. When Neil Armstrong reported his "giant leap for mankind" to Mission Control his words went immediately around the world and into history. The Apollo Mission Control Center and Launch Complex 39 at the Kennedy Space Center are the two resources that symbolize for most Americans achievements of the manned space program leading to the successful first moon landing during the flight of Apollo 11 in July 1969.

The MCC was designated a National Historic Landmark on October 3, 1985, and NASA essentially froze it in time, making it available for public tours through the Johnson Space Center Visitors' Center.

The case of the Lunar Landing Research Facility (LLRF), at Langley Research Center in Hampton, Virginia, is quite different. Built in 1965 at a cost of $3.5 million, it was a large A-frame structure, 400 feet by 230 feet, where a lunar landing training simulator allowed astronauts to practice landings in the simulated one-third gravity of the lunar surface. It also served as a moonwalking simulator for Apollo astronauts by suspending the subject so that he was free to generate walking movements on a plane inclined at 80.5 degrees. This, coupled with a trolley system, made it possible to simulate multiple gravitational fields. The structure remains an impressive site on the horizon near Hampton, where it can be seen above the trees from a great distance.

After Apollo, NASA maintained the LLRF intact but modified it for another purpose. It established there the Impact Dynamics Research Facility, a complex designation that meant that NASA used the A-frame as a place from which to drop aircraft for the study of impact on the ground at various inclinations, speeds, and the like. NASA removed a simulated lunar landscape that had previously existed and built a runway to simulate various crash environments. The LLRF was designated a National Historic Landmark on October 3, 1985, in no small measure because

experiences gained by the Apollo astronauts on the Lunar Landing Research Facility indicated that it was possible to successfully master the complicated skills that were required to land the LEM on the Moon. Both Neil Armstrong and Buzz Aldrin trained there for many hours. Only when they successfully mastered skills necessary to fly the LEM would NASA approve plans for their historic first landing on the Moon in July 1969.

Historic preservationists agreed that the LLRF was an indispensable tool that enabled NASA to land Americans on the Moon by July 1969.

Furthermore, NASA cast the history of the LLRF as critical to the Apollo effort. As it stated concerning the National Historic Landmark in its fact sheet on Langley Research Center's contributions to the Apollo program:

> Langley's Lunar Landing Research Facility, completed in 1965, helped to prepare the Apollo astronauts for the final 150 feet of their lunar landing mission by simulating both the lunar gravity environment and full-scale LEM vehicle dynamics. The builders of this unique facility effectively canceled all but one-sixth of Earth's gravitational force by using an overhead partial-suspension system that provided a lifting force by means of cables acting through the LEM's center of gravity.

Twenty-four astronauts practiced lunar landings at this facility, the base of which was modeled with fill dirt to resemble the surface of the Moon. Armstrong and Aldrin trained on it for many hours before liftoff of Apollo 11. As was the case with all space missions, the successful landings of the first two men on the Moon depended heavily on expert training in ground equipment like Langley's Rendezvous Docking Simulator and LLRF. With a sense of pride in the role this site played in one of humanity's greatest adventures—perhaps its greatest—NASA nonetheless was unwilling to ensure funding necessary for preservation of this site into the future.

The debate over the LLRF reached a crisis point in the mid-1980s when NASA leaders made clear to the National Park Service that the agency had no funds for upkeep of sites ancillary to the mission of the agency despite their historical significance. By that time, the crash test program had ended and NASA officials argued that preservation of a large open-air historic site presented numerous challenges for the space agency. NASA had long tried to gain an exemption from the requirements of the National Historic Preservation Act of 1966 for proper preservation of designated historic sites either on the National Register of Historic Places or as one of the select few National Historic Landmarks. Several of NASA's sites, including the LLRF, were in the latter category.

It has always been a challenge to balance historic preservation with reuse of facilities, but NASA began a campaign in the mid-1980s to enjoy the benefits of recognition without the requirements of maintaining facilities in line with preservation law. A study identified several sites for preservation; for each one, NASA tried to delay designation. During deliberations over designation of many of them in 1986, NASA's associate administrator for management stated:

This Agency has a dynamic research and development mission which requires that we make maximum use of limited Federal government resources. To do so, we are constantly modifying, rehabilitating, reconfiguring, adjusting, and altering our facilities to meet new program requirements . . . in the interest of minimizing Federal expenditures for facilities, we plan to continue to change these facilities as needed to meet future programs.

This correspondence did not result in the desired exemption and NASA made other entreaties to both the National Park Service and the Department of the Interior. On October 2, 1987, NASA Administrator James C. Fletcher told Secretary of the Interior Donald P. Hodel, "NASA simply cannot afford to become entangled in time consuming protracted negotiations over the status of planned changes in operational facilities which are absolutely crucial to the Nation's continuing aeronautics and space research, technology, and exploration missions. . . . Accordingly, I have no choice but to request that you take action to dedesignate the facilities (NASA NHLs) described in Enclosure 1 as historic landmarks." No action resulted. NASA made the same request in 1989, again with no resolution.

Prevailing on Rep. Robert Walker (R-PA), the ranking minority member on the Committee on Science, Space, and Technology, in 1989 NASA pursued a legislative waiver in the agency's fiscal year 1990 authorization bill to exempt its National Historic Landmarks from provisions of the National Historic Preservation Act of 1966. The preservation community responded by persuading Congress to delete this language from the NASA appropriations bill, but this did not resolve the issue. Other scientific organizations, especially the National Science Foundation, have also weighed in to obtain waivers from the legislation that governs the management of designated historic sites. As Harry Butowsky of the National Park Service suggested in 1990, as this debate was taking place,

the question that the listing of technological facilities in the National Register of Historic Places has raised is the general perception among members of the scientific community who fear that such a move would severely limit their ability to upgrade or modify their facilities. While the National Park Service continues to believe that the designation of properties as National Historic Landmarks and their listing in the National Register of Historic Places are compatible with their continuing function as scientific resources, members of the scientific community have expressed their concerns. During the next few months all of the interested parties must see if an agreement is possible that will satisfy the concerns of the National Science Foundation and the owners of the observatories so that both the historical significance of these properties can be recognized and important scientific research can continue as in the past.

The matter did not rest long like this. In 1990 Congress asked for an analysis of what should be done to reach some accommodation, and the result involved undertaking a study by the Advisory Council on Historic Preservation to set measures that presumably all could endorse. Instead, the recommendations of this group pushed back on the position of NASA's leadership. The Advisory Council turned the NASA position on its head: "Given the late 20th century's pattern of rapid technological change, however, the protection of the physical environment that facilitated that change takes on increased importance. Federal agencies managing or assisting scientific research have a leadership role in the stewardship of historic properties under NHPA." The study emphasized that it was incumbent on federal scientific organizations, funded as they were by tax dollars, "to consider the effects of their actions on the historic values embodied in select facilities." Its recommendations insisted that scientific and technical agencies be allowed flexibility in the planning and execution of research work in their facilities but also serve as stewards over the nation's scientific heritage.

It also noted that the historic preservation community is largely unfamiliar with the scientific and technical nature of historically significant properties, while the technical overlords of these properties are unfamiliar with historical preservation requirements. Both would do well to engage the other in reaching accommodations. While this discussion did not end the dispute, NASA thereafter has largely been allowed to make changes to its National Historic Landmarks without much oversight beyond photographic documentation.

This situation, specifically in the context of the LLRF, came to a head in August 2004 when NASA proposed demolishing the structure, as well as several other historic properties at the Langley Research Center, because they had no future use. While the action has not yet been undertaken, NASA intends to document the facilities, submit reports through the Historic American Buildings Survey and the Historic American Engineering Record, and only then undertake demolition. In the process, an important part of America's Apollo material culture heritage will be lost forever. At the same time, without a use for the site, NASA officials are understandably hesitant to maintain it indefinitely. Clearly, the space agency is not responsible for historical preservation and its leaders have not agreed to preserve such sites as a part of its ongoing mission. There is no satisfactory solution that anyone can envision for this important site in the 21st century.

In the aftermath of the Apollo program, NASA transferred to the Smithsonian Institution's National Air and Space Museum (NASM) all flown Apollo command modules. The museum has placed them on display in various museums around the nation, as well as at the London Science Museum. In 1966 leaders of NASA and NASM negotiated an agreement that allowed NASM a right of first refusal to acquire artifacts deemed to be of historic significance, once they were declared surplus by NASA and offered for disposition through the federal property system. This agreement has been in effect since that time. Accordingly, NASA transferred to NASM its entire collection of flown Apollo spacecraft following the conclusion of the program, and as NASA made clear, it did so because of the costs required to keep them in-house for the purposes of historic preservation. Since that time, the Smithsonian has managed them.

The centerpiece of this historic display over the years has been the Apollo 11 command module, on display at the National Air and Space Museum. Sent on a 50-state tour before reaching the museum, it was unveiled in a ceremony on the fifth anniversary of the first Moon landing. Apollo 11 immediately became a favorite for visitors, with journalists using it as punch line for a renewed sense of space exploration. One editorialized, "It would be an enormous folly if we remain content to see Neil Armstrong's 'giant leap for mankind' carry us no further than the floor of the Smithsonian." The visitors to the National Air and Space Museum are certainly there to see what most consider wonders of modern technology, and at some level this capsule is a shrine to American greatness, treated as such in the museum. All Apollo capsules, wherever they might be displayed, have become tangible expressions of an American sense of mastery. They are the modern equivalent of the pyramids.

Finally, the six Apollo landing sites on the Moon pose an interesting problem for the material culture of the Apollo program in modern society. Since these missions were completed in 1972, only a few robotic missions by the Soviet Union and one by China in 2013 have landed on the Moon. None of those missions has disturbed those sites. That does not mean that no one will do so in the future, once humankind goes back to the Moon. The people and rovers that return might not be American, though; several nations have hinted at long-term plans to explore the Moon, and several commercial ventures have been developing robotic rovers that might be sent there. This issue had arisen as early as 1971, when the president received an offer from a private citizen to buy artifacts left by the Apollo astronauts on the Moon; NASA chose not to accept this offer.

Since that time, the issue of how to preserve these artifacts has arisen repeatedly. In 1984, NASM Director Walter J. Boyne formally asked that NASA transfer the Apollo lunar surface objects to the Smithsonian Institution to become part of the National Collection, remarking:

> Although there are, at present, no plans for return visits to the moon, it is certain as anything that someday man will return. When he does, it is imperative that the historic and scientific significance of all remnants of earlier, pioneering exploration efforts be fully appreciated and respected. We believe there is no better way to guarantee that the items will be preserved for appropriate scientific, historic and educational used than to have the items registered in advance as belonging to the National Collection of Space Artifacts.

NASA and NASM staff worked together to identify the objects left at each of the Apollo landing sites and over the course of the next two years came up with a basic list. After all of this work, both NASA and NASM quietly dropped the matter, and nothing more came of it. In part this was because of the all-consuming nature of the *Challenger* accident and NASA's recovery from it in the 1986–88 period and Boyne's resignation as NASM director in August 1986. Neither organization after 1986 had senior officials seeking to bring this transition to completion.

For the last two decades, private citizens and organizations have advocated preservation of the Apollo landing locations as historic sites. This is very much part of the wishful thinking surrounding the prospect of space tourism. Television personality Lou Dobbs commented on this possible commercial activity: "There is also the bastard child of space business—tourism. Long scoffed at by serious space explorers, space tourism could actually become one of the driving financial forces of s-commerce." Like many things in the space business, however, space tourism always seems to be a decade away. Even so, we must plan for the eventuality. Since China has now become a spacefaring nation, there have been rumors that it intends to extend its human presence to the Moon's surface. Additionally, India plans to land robots there. Once tourists reach the Moon, they will immediately set out for the Apollo landing sites, and although that may be many years in the future, they will be a popular tourist destination.

In 1999, New Mexico State University, working with the Lunar Legacy Project and using $23,000 in funding from NASA's New Mexico Space Grant Consortium, prepared a nomination for Tranquility Base, the Apollo 11 landing site, to be designated a National Historic Landmark. The project leaders wrote:

> Although this site is not yet 50 years in age, we believe its
> overwhelming significance makes it eligible for such a nomination.
> The first lunar landing site is not on United State government
> property, nor is it on property controlled or leased by the United
> States. The first lunar landing site is on "neutral territory" in space,
> but technically it is under the jurisdiction of the U.S. According
> to "The Treaty on Principles Governing the Activities of States in
> the Exploration and Use of Outer Space, Including the Moon and

Other Celestial Bodies" (1967 signed by the U.S.) a country which launches any objects into space retains possession and control of the objects indefinitely. Therefore, the United States retains possession and control of all objects it has placed on the moon. The Apollo 11 Eagle Landing Pod, the United States Flag, the two scientific objects which make up the district of structures on Tranquility Base are still possessions of the United States.

Based on this rationale, the proposal argued for recognition through the National Historic Preservation Act of 1966.

This request was not acted upon. That decision—or indecision—led to a request from Ralph D. Gibson Jr., an anthropology graduate student at New Mexico State University who had been working on the proposal, to President Bill Clinton in 2000 to intercede with the National Park Service. The Park Service responded: "It has been determined as a matter of policy that it would not be appropriate to designate National Historic Landmarks on the Moon." Scratched out of the draft of this letter was a statement: "The Moon is not territory that belongs to the United States." This deletion may have been made because of the National Historic Landmark status of sites in Morocco and the Republic of Palau that the United States did not technically own. The letter also noted that the National Park Service did not believe it could exercise the jurisdiction over lunar sites necessary to maintain their integrity. Others have suggested since 2000 that these sites should be designated under United Nations authority as World Heritage Sites to be protected for all time, and they have petitioned the leadership of the United States to make this a priority at the UN. Others have suggested that the Smithsonian establish a bureau on the Moon to display the objects left there by the Apollo astronauts. To date, nothing has been done.

At present, the historic sites on the lunar surface are pristine, frozen in time like the Apollo events that occurred there. I remember how Ron Nelson, former director of the Bishop Hill State Historic Park in Illinois, used to describe the perfect historic site: one without any visitors. The presence of visitors, even in small numbers and tightly controlled, could subvert the historicity of any site. But it is unlikely to remain that way indefinitely. Few take planning for the preservation of the Apollo sites on the Moon seriously, but this issue has arisen with greater immediacy as time has progressed. For instance, in 2007, the X-Prize Foundation issued a call for contestants to participate in privately funded missions to send a robot to the lunar surface and send back to Earth specific data and imagery, among them pictures of Apollo landing sites. The winner would receive a $20 million base prize, with the potential for further funding if it accomplished additional tasks on the Moon. As the teams vying for the prize demonstrated advancing capabilities, NASA worked to ensure the integrity of the Apollo sites. Not initially motivated by concerns with historic preservation—scientific and technical preservation issues dominated—NASA did agree that all must work to preserve this history on the lunar surface. Accordingly, in February 2011, it said that it wanted "to establish special status for the Apollo 11 sites with a broad keep out zone recommended around the site. The remainder of the sites where NASA spacecraft have landed or crashed can be visited with smaller keep out zones identified so that valuable data at these sites can be collected without compromising the site." This initial statement of NASA's position is a major step forward in preserving Apollo and other sites on the Moon, but there are now no laws in place that govern the conduct of any entity that might visit these sites on the Moon.

Many people believe that it is now time for more thorough protocols to be developed and implemented for preserving these historic lunar sites. A major step forward came in July 2011 when NASA issued guidelines that prescribed

> approaching Apollo landing sites and artifacts at a tangent, to avoid crashing into them, and suggests no-fly and buffer zones to avoid spraying rocket exhaust or dust onto historic equipment. The document also includes a research wish list, written by NASA scientists and engineers, for any private team, or country, sending a craft to the moon. The list ranges from the mundane, such as taking close-up photographs of decades-old laser range-finding mirrors still used by Earth-based astronomers, to more far-out ideas, such as studying discarded food or abandoned astronaut feces.

These are not laws, but they do enumerate some standards of conduct for any mission going to these lunar sites. I have joked for years that many museum professionals, myself included, would be pleased to fly the mission to the Moon to put ropes and stanchions around these sites for visitor flow control. Absent such a mission, these guidelines are a useful plan B. Clearly, what should not happen is what has taken place in Antarctica, where a failure to undertake proper preservation has led to visitors vandalizing and ransacking sites of rich historic interest established by explorers.

More recent efforts have somewhat muddied the waters. In July 2013, Rep. Donna Edwards (D-MD) placed in consideration by the House of Representatives H.R. 2617, the Apollo Lunar Landing Legacy Act. Had it been enacted, this act would have designated the Apollo landing sites as a national park under the jurisdiction of the US Department of the Interior. Protagonists found many of the provisions of this act detrimental to long-term space operations and recommended instead an international agreement on the preservation of lunar artifacts among the United States, Russia, and China. They argue that this approach "would be a far superior and long-lasting solution than the unilateral US proclamation in H.R. 2617. Enforcement of the agreement would be through each nation's national laws, applying to those entities subject to the jurisdiction or control of the agreement members. Each nation's property would be protected and preserved." The issue remains unresolved.

These six episodes in the preservation of the material culture of the American race to the Moon offer very different experiences. There is no question that these various sites are historically significant, perhaps as significant as any human site on Earth. Ensuring that some future nation, company, or person does not destroy the heritage of Apollo, as has been done repeatedly in many other settings, is an important goal. We do not want to allow a repeat of Napoleon in Egypt, or the results of the American invasion of Baghdad in 2003, or the loss of Antarctic history from tourists and others pillaging historic camps on the ice, to name only three instances.

Preserving the Apollo experience perhaps will become increasingly important the farther we move from that one brief time in history, a kind of Camelot in which all the cosmic tumblers clicked into place to make possible successful Moon landings.

10

Denying the Apollo Moon Landings

For many years, when the July anniversary of Apollo 11 arrived, I gave public presentations at the Smithsonian Institution's National Air and Space Museum in Washington, DC, on the belief held by some people that humans never landed on the Moon. It was always an enormously popular event, attracting a large audience, some of them Moon-landing deniers; one presentation was filmed and is available on YouTube, where it enjoys a large viewership and draws a lot of weird comments. At one presentation, a high school teacher who firmly believed no one had ever been to the Moon challenged me, stating that he also believed no humans had ever flown in space. I was surprised and troubled at the thought that this person was out there shaping young minds. I urged skepticism of every claim, analysis of every question, and a willingness to follow data where it led without prejudice. My comments never persuaded him.

This story points up the thorny issues of belief, knowledge, rational thought, and myth in our recollections of the Apollo story. Most have celebrated the Moon landings as marking an American epoch. Others have criticized it as a waste either from the political left or the right. A very few—but a vocal few—deny that it happened altogether. Those who deny it do not accept the same rules of investigation and knowledge that all others live by. Instead, they are decidedly postmodern in the sense that evidence, for them, is situational and knowledge a commodity to be bought and sold.

Some antecedents to this position were seen almost from the point of the first Apollo missions, when a small group of Americans denied that they had taken place at all. It had been faked in Hollywood by the federal government, they argued, for purposes ranging—depending on the particular Apollo landing denier—from embezzlement of the public treasury to complex conspiracy theories involving international intrigue and

The Apollo 14 crew train for their lunar mission on December 8, 1970. Conspiracy theorists have used images such as this one to argue that the United States never landed on the Moon. (NASA image no. 70P-0503)

murderous criminality. The deniers tapped into a rich vein of distrust of government, populist critiques of society, and questions about the fundamentals of epistemology and knowledge creation. They were a tiny lot. At the time of the first landings, opinion polls showed that overall less than 5 percent of Americans "doubted the moon voyage had taken place."

Apollo 8 astronaut Bill Anders thought that live television would help convince skeptics, since watching "three men floating inside a spaceship was as close to proof as they might get." He could not have been more wrong. Fueled by conspiracy theorists of all stripes, this number has grown over time. In a 2004 poll, while overall numbers remained about the same, among Americans between 18 and 24 years old, "27%

expressed doubts that NASA went to the Moon," according to pollster Mary Lynne Dittmar. Doubt is different from denial, but it was a trend that seemed to be growing over time among those who did not witness the events.

Americans, certainly, and perhaps all the cultures of the world, love the idea of conspiracy as an explanation of how and why many events have happened. It plays to their innermost fears and hostilities to think that there is a well-organized, well-financed, and Machiavellian design being executed by some malevolent group, the dehumanized "them," who seek to rob "us" of something we hold dear. One scholar defined a conspiracy, and this represents a practicable approach to the topic, as "the attribution of deliberate agency to something that is more likely to be accidental or unintended." This certainly happens often enough. And in all cases these tend to be exaggerated, expanded, and embellished with every retelling.

Conspiracy theories abound in American history. Oliver Stone's film *J.F.K.*, presenting a truly warped picture of recent American history, shows how receptive Americans are to believing that Kennedy was killed as a result of a massive conspiracy variously involving Cuban strongman Fidel Castro, American senior intelligence and law enforcement officers, high communist leaders in the Soviet Union, union organizers, organized crime, and perhaps even Vice President Lyndon B. Johnson. Stone's film brought the assassination conspiracy to a broad American public. For years amateur and not-so-amateur researchers have been churning out books and articles about the Kennedy assassination conspiracy—more than 5,000 of them. It has been one of the significant growth industries in American history during the last half-century. These efforts, especially the Oliver Stone film, even prompted the US Congress to pass the President John F. Kennedy Assassination Records Collection Act of 1992, making available huge amounts of government documents, audiotapes of witness interviews and phone calls, and other information on the assassination of Kennedy.

Lest you think these are diversions for those who have nothing better to do or that they are the hobgoblins of cracked minds, some conspiracies have been instrumental in charting major turns in the direction of the nation. The most striking example is the American Revolution. When the British Empire finally defeated France in the Seven Years' War in 1763,

Great Britain turned its attention to its colonies as it had not done before, in part to exact taxes from them to help pay for the war and the other costs of empire. The Sugar Act, the Townshend Duties, the Stamp Act, the Intolerable Acts, the Quebec Act, and a host of other laws designed to raise revenue riled American colonists to rebellion. In some respects, the United States was born out of a tax revolt, and one can only imagine what would have been Sam Adams's and Thomas Jefferson's reaction had the British tried to impose an income tax.

Taken together, these and other efforts by the British government were portrayed by rebellious colonists as a conspiracy to rob Americans of their rights as Englishmen (women did not even enter into the picture at that point). Ultimately, colonials argued that a grand conspiracy to enslave Americans was underway and that they were compelled to stand together to defend their liberties and defeat a determined, evil oppressor. Interestingly, the liberty/slavery rhetoric had the potential to enflame many Americans, since they saw the dichotomy between freedom and slavery every day in the colonies' cities and especially on the plantations. A conspiracy to enslave white Americans, therefore, was an especially potent force in motivating revolution.

Numerous other instances of significant movements in American history have been motivated at least in part by the possibility of conspiracy. The terror attacks on September 11, 2001, moved conspiracy theories to the center of American life, as all manner of conjecture emerged about the attacks, virtually all of it easily proven false and dispatched as groundless. Still, the rumors and theories evolved with every retelling into ever more complex and outrageous stories. The 9/11 "truth movement," fed by the Internet, saw conspiracy theorists debating whether the government had allowed or even fomented the attacks to gain political advantage. Some on the far right likewise have denied the school shootings of the past decade ever happened, explaining them away as theater staged to create an environment whereby guns might be outlawed in the United States. In the period since the presidential election of 2016, conspiracy theories about a "Deep State" have been used to delegitimize the functioning of government organizations.

In the case of the Moon-landing deniers, the interrelationships of four key elements—dualism, scapegoating, demonization, and apocalyptic aggression—are linked in the utter disbelief, distrust, and apparent hatred of anyone who suggests that their presumed evidence of a massive government conspiracy is unpersuasive. I would add that their so-called evidence is outlandish and unworthy of receiving any credence. Historian David Aaronovitch commented about these arguments, "It offended my sense of plausibility," adding,

> My uncogitated objection ran something like this. A hoax on such a grand scale would necessarily involve hundreds if not thousands of participants. There would be those who has planned it all in some Washington office; those in NASA who had agreed; the astronauts themselves, who would have been required to continue with the hoax for the whole of their lives, afraid even of disclosing something to their most intimate friends at the most intimate moments; the set of designers, the photographers, the props department, the security men, the navy people who pretended to fish the returning spacemen out of the ocean and many, many more. It was pretty much impossible for such an operation to be mounted and kept secret, and inconceivable that anybody in power would actually take the risk that it might be blown.

That is the reaction of most observers who hear the argument that Apollo astronauts never landed on the Moon. The conspiracy theory—or more appropriately *theories*, since every adherent has his own, and all seemingly compete with one another in complexity and lack of verisimilitude—is attractive to those wanting to disbelieve claims of authority figures. Some who know better invoke it in passing as a joke, but those who hold to the conspiracy framework often possess a deep skepticism, even a resentment, of national authority.

Since before the conclusion of the Apollo program in 1972, some have questioned the Moon landings, claiming that NASA faked them in some way or another, presumably with the acquiescence or perhaps even active involvement of other individuals and organizations. Some of those skeptical of the Apollo flights made their cases based on naïve and poorly constructed knowledge, but imagery from space did not seem to convince them of any reality beyond what they already wanted to believe. For example, my paternal grandfather, Jeffrey Hilliard Launius, was a 75-year-old farmer from southern Illinois at the time of the first Moon landing in 1969. A Democrat since the Great Depression of the 1930s—because, as he said, Roosevelt gave him a job with the WPA when he could not feed his family and was on the verge of losing everything—his denial of the Moon landing was based essentially on lack of knowledge and naïveté. In his estimation such a technological feat was simply not possible. Caught up in the excitement of Apollo 11 in the summer of 1969, I could not understand my grandfather's denial of what appeared obvious to me. He did not assign any conspiratorial motives to the government, especially the Democrats; after all, he had trusted the party implicitly for more than 35 years. Even now, I still cannot fully fathom his conflicting positions—trust of the Democrats in government and unwillingness to believe what they said about the Moon landing. In his insular world, change came grudgingly, however, and a Moon landing was certainly a major change. As a measure of his unwillingness to embrace change, my grandfather farmed his entire life with horses rather than adopting the tractor because, in his estimation, tractors were "a passing fad." At the time of his death, in 1984, Jeff Launius still did not believe that Americans had landed on the Moon.

President Bill Clinton recalled in his 2004 autobiography a similar story of a carpenter he worked with in August 1969, not long after the Apollo 11 landing:

> Just a month before, Apollo 11 astronauts Buzz Aldrin and Neil
> Armstrong had left their colleague, Michael Collins, aboard
> spaceship Columbia and walked on the Moon, beating by five
> months President Kennedy's goal of putting a man on the Moon
> before the decade was out. The old carpenter asked me if I
> really believed it happened. I said sure, I saw it on television. He
> disagreed; he said that he didn't believe it for a minute, that "them
> television fellers" could make things look real that weren't.

Clinton thought him a crank at the time and afterward a homespun skeptic. He then allowed that a healthy criticism of everything was not necessarily a bad idea.

How widespread were the skeptics about the Moon landings in the 1960s? That is almost impossible to say. For example, *New York Times* science reporter John Noble Wilford remarked in December 1969 that "a few stool-warmers in Chicago bars are on record as suggesting that the Apollo 11 moon walk last July was actually staged by Hollywood on a Nevada desert." More important, the *Atlanta Constitution* led a story on June 15, 1970, with "Many skeptics feel moon explorer Neil Armstrong took his 'giant step for mankind' somewhere in Arizona." It based its conclusion that an unspecified "many" questioned the Apollo 11 and 12 landings, and presumably the April 1970 accident aboard Apollo 13, on an admittedly unscientific poll conducted by the Knight newspaper chain of 1,721 US citizens in "Miami, Philadelphia, Akron, Ohio, Detroit, Washington, Macon, Ga., and several rural communities in North and South Carolina." Those polled were asked, "Do you really, completely believe that the United States has actually landed men on the moon and returned them to earth again?" While numbers questioning the Moon landing in Detroit, Miami, and Akron averaged less than 5 percent, among African Americans in such places as Washington, DC, a whopping 54 percent doubted the Moon voyages. That perhaps said more about the disconnectedness of minority communities from the Apollo effort and the nation's overarching racism than anything else. As the story reported, "A woman in Macon said she knows she couldn't watch a telecast from the moon because her set wouldn't even pick up New York stations."

Not everyone who denied the Moon landings at the time was so naïve and simplistic in their assessments. Some spun conspiracy theories of complex structure and shocking intent. As Howard McCurdy opined, "To some, the thrill of space can't hold a candle to the thrill of conspiracy." Over the years many conspiracy scenarios have been concocted, and it sometimes appears that the various theorists are even more cantankerous toward rival theories than they are toward NASA and the Apollo program. An early and persistent theme has been that as a Cold War measure the United States could not afford to lose the race to the Moon, but when failure loomed NASA faked the landing to save face and national prestige. It used the massive funds dedicated to the effort to "pay off" those who might be persuaded to tell the truth; it also used threats and in some instances criminal actions to stop those who might blow the whistle. One of the most common assertions has been that in the latter 1960s, the US government was in disarray because of the debacle of the Vietnam War, the racial crisis in the cities, and social upheaval. The Apollo program proved an ideal positive distraction from this strife, a convenient conspiracy designed to obscure other issues. One story published in 1970 stated this belief as expressed by an African American preacher: "It's all a deliberate effort to mask problems at home," *Newsweek* quoted him saying. "The people are unhappy—and this takes their minds off their problems."

Other conspiracies were more absurd. For example, William Brian asserted that perhaps Americans did go to the Moon, but they did so through the means of some extraterrestrial technology. In his estimation, NASA employed captured—or perhaps given—technology from beings beyond Earth to reach the Moon. This forced the agency to create a cover story for more sinister purposes. "You can't let one bit of information out without blowing the whole thing," he noted. "They'd have to explain the propulsion technique that got them there, so they'd have to divulge their UFO research. And if they could tap this energy, that would imply the oil cartels are at risk, and the very structure of our world economy could collapse. They didn't want to run that risk."

The first conspiracy theorist to make a sustained case for denying that the United States landed on the Moon was Bill Kaysing, a journalist who had been employed for a few years in the public relations office at Rocketdyne, Inc., a NASA contractor, in the early 1960s. His 1974 pamphlet *We Never Went to the Moon* laid out many of the major arguments that have been followed by other conspiracy theorists since that time. His rationale for questioning the Apollo Moon landings offered poorly developed logic, sloppily analyzed data, and sophomorically argued assertions. Kaysing believed that a failure to land on the Moon sprang from the idea that NASA lacked the technical expertise to accomplish the task, requiring the creation of a massive cover-up to hide that fact. He cited as evidence perceived optical anomalies in some imagery from the Apollo program, questioned the physical features of certain objects in the photographs (such as a lack of a star field in the background of lunar surface imagery and a presumed waving of the US flag in an airless environment), and challenged the possibility of NASA astronauts' surviving a trip to the Moon because of radiation exposure.

Others followed in Kaysing's footsteps, arguing one conspiracy or another, none with compelling evidence and often with nothing that might be considered anything more than assertions. As John Schwartz wrote of the conspiracy theorists in the *New York Times* on July 13, 2009, "They examine photos from the missions for signs of studio fakery, and claim to be able to tell that the American flag was waving in what was supposed to be the vacuum of space. They overstate the health risks of traveling through the radiation belts that girdle our planet; they understate the technological prowess of the American space program; and they cry murder behind every death in the program, linking them to an overall conspiracy."

Ted Goertzel, a professor of sociology at Rutgers University who has studied conspiracy theorists, told Schwartz that "there's a similar kind of logic behind all of these groups." For the most part, he explained, "They don't undertake to prove that their view is true" so much as to "find flaws in what the other side is saying." And so, he said, argument is a matter of accumulation instead of persuasion. "They feel if they've got more facts than the other side, that proves they're right."

The nature and extent of the Moon-landing conspiracy has differed from advocate to advocate with reckless abandon. They seemed to take one of three general paths. The first, and the most significant, was that the entire human landing program for the Moon was a sham, manufactured for public consumption by an evil political establishment. According to the conspiracy theorists NASA undertook this elaborate ruse for several reasons. The US government felt pressure to win the space race against the Soviet Union, but with technical challenges too great to overcome NASA decided on pursuing an elaborate sham. Bill Kaysing made this case in his 1974 pamphlet. He insisted that even though the Soviet Union was watching the American effort closely, without any evidence whatsoever, it was easier to successfully fake it than actually to land on the Moon. He even speculated that the chance of landing successfully on the Moon stood at 0.017; on what this calculation was based is a mystery and does not square with NASA estimates at the time, which stood at approximately 87 percent for at least one successful landing before the end of the 1960s. In addition, Moon-landing deniers asserted that the Apollo program served as a distraction from the other problems of the United States during the 1960s, especially the Vietnam War, despite the fact that it was not a particularly successful distraction judging from the sustained opposition of the American public.

In addition to the full hoax argument, as a second approach, some Moon-landing deniers concede that there were robotic missions to the Moon, but that the human Apollo landings were faked. Professional Moon-landing denier Bart Sibrel, in particular, has asserted that Apollo spacecraft crews had faked their orbit around the Moon and their walk on its surface by using trick photography, but did accept Earth orbital missions by them. They could not go to the Moon, Sibrel and deniers of his ilk claim, because going beyond the Van Allen radiation belts would have given them lethal doses of cosmic radiation. A few conspiracy theorists in this category even allow that NASA landed robotically on the Moon various passive reflector mirrors used for laser ranging and other human-made objects to bamboozle the public. Then, third, there are those who believe that humans went to the Moon, but did so with the assistance of extraterrestrial visitors—or that Apollo astronauts discovered extraterrestrial life there. These claims ranged from gravitational anomalies to alien artifacts to alien encounters. Accordingly, this brand of conspiracy theorist claim that NASA covered up what had been found, in the manner the discovery of a monolith at Clavius Crater on the Moon in *2001: A Space Odyssey.*

For example, Richard Hoagland has asserted for many years that the Apollo program discovered large artificial glass structures on the lunar surface that has been kept from the public. Besides other conventions common in a cover-up, Hoagland made the claim that the astronauts that went to the Moon had been hypnotized and any memories of extraterrestrial encounters were removed. Most interestingly, Hoagland has argued that NASA deviously orchestrated the origins of the Moon-landing denials as a disinformation campaign to mask the discovery of extraterrestrial structures on the lunar surface. As recently as September 25, 2009, Hoagland asserted that the water molecules that NASA's Lunar Reconnaissance Orbiter had discovered on the Moon obviously had been leaked from buried extraterrestrial cities. No evidence supported these assertions.

While the various claims of the Moon-landing deniers have evolved over time, their reasons for making these claims have rested on five types of "evidence." First and by far the most significant type are anomalies found in photographs or, to a much lesser degree, movies taken on the missions. Of course, for all but a handful of American astronauts, the voyages of exploration to the Moon during Project Apollo were events that Earthlings participated in vicariously from more than 243,000 miles away. In such a setting, therefore, imagery has played a critical role in the communication of the experience. At some level, the questioning of this imagery as a witness to the Moon landings represented a direct assault on the episode in human history. While Apollo imagery documented in graphic detail what took place on the Moon, the use of that same imagery to raise questions about the entire enterprise is an irony too great to ignore. Such a trend has also been common in virtually every contested aspect of history since the rise of photography in the first half of the 19th century. It has been present in efforts to deny the Holocaust, the single-shooter theory of the Kennedy assassination, and the official explanation of the 9/11 attack, so its presence as the centerpiece of the Moon-landing denials should not be surprising.

At the same time, only some 25 images have been invoked in such a set of claims. These include images that do not show stars in the background, despite conspiracy theorists' insistence that they should be clearly seen; the fact that dust was not present on the landing pads of the spacecraft; the assertion that shadows and lighting on the Moon are uneven and counterintuitive to the photographs in which they are seen; that flags seem to be blowing in a breeze although there is no wind on the Moon; that some rocks appear to have propmaster marks on them; and that Réseau-plate crosshairs sometimes seem to disappear behind objects in an image. For each of these charges, there are completely reasonable, understandable, and convincing explanations, most relating to the nature of photography and the vicissitudes of shadows, lighting, and exposure of film in a vacuum. Few Moon-landing deniers, however, will accept any explanation whatsoever. I have personally had one say to me, "Regardless of whatever you might say, I will never believe that humans have landed on the Moon."

The second major type of "evidence" offered by Moon-landing deniers relates to radiation and its survivability by astronauts in space, especially radiation from the Van Allen radiation belts. While there is indeed radiation both in the Van Allen belts and beyond, and radiation's risks to human health are real, contentions that it would not be survivable are at least sophomoric and perhaps intentionally misleading. All those engaged in spaceflight, from its beginning to the present, are concerned with challenges of radiation in safely undertaking flights to the Moon, as well as in Earth orbit, but radiation is certainly manageable. They recognized that the risk was dependent on two factors: the amount of radiation received and the length of time to which astronauts were exposed to it. In both cases, these were minimal. In the end, the total radiation dose received by each of the astronauts was about one rem, while radiation sickness does not take place until one receives a dose of 100 to 200 rem, and fatal dosages are somewhere in the 300-plus-rem range. Future long-duration deep-space missions remain a concern for flight surgeons and life scientists, but countermeasures to the effects of radiation have made rapid progress in recent decades, and those are much more sophisticated than lead shielding.

Third, the Moon-landing deniers have claimed that the technology necessary to accomplish the Apollo program did not exist at the time, and that it could not have been created in the time necessary to meet the end-of-decade landing commitment. Moreover, they argue that the United States, 50 years after Apollo 11, still does not possess the technology necessary to go the Moon. These claims are statements of either remarkable naïveté or obtuseness. For example, Apollo landing denier Bart Sibrel has claimed:

Almost 40 years ago, with *combined* CSM and LM guidance computer memory totaling only 10.3% [152kb] of a common 1.4MB [1474.56kb] floppy disk [emphasis in original], NASA claims to have traveled 60,000% as far as any other manned spacecraft has gone *before or since*. Basically a household calculator (or discount watch) took 27 men [Apollo 8 to 17] to the moon and back, with the help of slide rules—accounting for fuel consumption, angle of approach, lunar landing, rate of descent, and so on. . . . These limitations alone, make the trip to the moon a theory, and *not* a fact.

Sibrel fails to acknowledge that thousands upon thousands of individuals supported the Apollo missions to the Moon with banks of ground-based computers and the collective brainpower of engineers with mechanical calculators and slide rules, both in detailed planning and throughout the missions. To deny that human beings with less technology than presently available could manage the missions grotesquely underestimates human ingenuity. It is a bit like the assertions of Erich von Däniken in the 1960s, who insisted that extraterrestrials with interstellar technology had to have built the pyramids and other ancient wonders because Earthlings never could have done so using only ingenuity, simple machinery, and muscle power. It was disingenuous then, and it remains so now.

The technologies necessary for the Moon landings did not exist at the outset of the program, true, but those technologies—while sophisticated and impressive—were largely within the grasp of the United States at the time of the 1961 decision. More difficult was ensuring that those technological skills were properly managed and used. The rise of the program management concept, configuration control, and systems integration were critical to successfully reaching the Moon by the end of the 1960s. At the same time, NASA experienced all manner of difficulties and some serious failures—costing the lives of three astronauts and nearly taking the lives of the Apollo 13 crew—throughout the program. The ancillary assertion that the United States cannot return to the Moon at present is a true statement, but it is also beside the point, since the hardware required for a Moon landing would have to be recreated using modern technology based on what is already known from the first experience and the evolutionary development of space technology since.

Fourth, Moon-landing deniers attack the physical evidence of the Moon rocks returned by the Apollo astronauts as either faked by NASA or collected in Antarctica prior to the missions. Those who claim that NASA faked the rocks early emerged but have been displaced by a more sophisticated but still far-fetched denial. For instance, some recent landing deniers claim that when NASA's Marshall Space Flight Center director, Wernher von Braun, visited Antarctica in 1967, it was so that he could explore the possibility of using meteorites from the Moon collected in the ice as stand-ins for the rocks claimed to have been from the Apollo landings. Antarctica has been a source of meteorites that emerge from ice as it shifts, and each Antarctic summer the National Science Foundation sends teams to the ice in search of new specimens. The six Apollo landing missions collected a total of 840 pounds of Moon rocks, and scientists worldwide have analyzed these at length. All agree that these rocks are from the Moon, and no papers in peer-reviewed scientific journals exist disputing the claim. This is not surprising, Moon-hoax advocates insist, since they are in actuality meteorites found in Antarctica. The one overriding problem with this theory is that the first lunar meteorite discovered in Antarctica was collected in 1979, a decade after the first Moon landing, and scientists did not realize its lunar origin until 1982 after comparing it to lunar samples acquired by Apollo astronauts. As one scientist stated: "About 36 lunar meteorites have been found in cold and hot deserts since the first one was found in 1979 in Antarctica. All are random samples ejected from unknown locations on the Moon by meteoroid impacts." Nothing approaching 840 pounds of lunar meteorites has been discovered, and this rarity means that this explanation is completely implausible.

Deniers of the Moon landing also point to anomalies in the historical record to cast doubt on the NASA account of the Apollo program. For example, one of the persistent beliefs is that the "blueprints" for the Apollo spacecraft and Saturn V rocket have been lost, or perhaps they never existed. This is viewed as evidence of a giant hoax on the part of NASA to fake the Moon landings. Aside from the question as to how such a loss might support NASA's culpability in a hoax, presumably their loss implies they never existed. This is simply untrue. The National Archives and Records Administration maintains a regional Federal Records Center at Ellenwood, Georgia, just outside of Atlanta, where the records from the Marshall Space Flight Center are housed. Those records include more than 2,900 linear feet of Saturn V records, including drawings and schematics. As Paul Shawcross, of NASA's Office of Inspector General, said in 2000, "The problem in recreating the Saturn 5 is not finding the drawings, it is finding vendors who can supply mid-1960s vintage hardware, and the fact that the launch pads and vehicle assembly buildings have been converted to space shuttle use, so you have no place to launch from. By the time you redesign to accommodate available hardware and re-modify the launch pads, you may as well have started from scratch with a clean sheet design." A similar story about the loss of the original broadcast video from the Apollo 11 landing has been used to cast doubt on the whole endeavor, causing NASA to undertake an unprecedented search for the tapes, finding some but not all that were missing.

Finally, in this same category of anomalies in the historical record, conspiracy theorists have scrutinized every word uttered by the Apollo astronauts over the years to try to catch them in some statement that might be interpreted as denying the landings. Having found none, some have resorted to selectively excerpting them and in some cases to making them up altogether. The classic example is Neil Armstrong, the first human to set foot on the Moon in 1969. Bart Sibrel commented, "Neil Armstrong, the first man to supposedly walk on the moon, refuses to give interviews to anyone on the subject. 'Ask me no questions, and I'll tell you no lies.' Collins also refuses to be interviewed. Aldrin, who granted an interview, threatened to sue us if we showed it to anyone." The implication is clear: They have something to hide and are unwilling to lie about a faked Moon landing.

Sibrel has gone further, accosting astronauts and demanding that they swear on the Bible that they walked on the Moon. Some have done so; others refuse to engage him. In one incident on September 9, 2002, Sibrel confronted Buzz Aldrin at a Los Angeles hotel and called him a "liar, a thief, and a coward." At that point Aldrin, then 72 years old, hit Sibrel with a right hook that sent him to his knees. While Sibrel pressed charges, the Los Angeles County District Attorney's office declined to pursue the incident. Most people who viewed video of this altercation expressed concern that Aldrin might have hurt his hand.

In a truly bizarre turn of events, on August 31, 2009, near the 40th anniversary of the first Moon landing, the satirical publication *The Onion* published a story entitled "Conspiracy Theorist Convinces Neil Armstrong Moon Landing Was Faked." The story had worldwide implications. This periodical is known for its outrageous humor, but this story was picked up as true in several newspapers around the world. Two Bangladeshi newspapers, the *Daily Manab Zamin* and *New Nation*, apologized afterward for reporting it as fact. It is bad enough when conspiracy theorists state such things without foundation; these then get picked up and broadcast by individuals, but when legitimate news organizations do so it is much more disturbing. As one commentator about this incident suggested, "Their excuse: 'We thought it was true so we printed it without checking.'" The writer added sarcastically. "Frankly, I understand them. After all, if it is on the internet, it *must* be real."

As it has turned out, throughout the latter third of the 20th century and into the 21st, with public confidence in the US government declining because of Vietnam, Watergate, and other scandals and malfeasance, it became somewhat easier for people to believe the worst about such a cover-up. For example, responding to a public opinion survey in 1964, 76 percent of the Americans polled expressed confidence in the ability of their national government "to do what is right" most or all of the time. This was an all-time high in the history of polling, and this goodwill helped lay the foundation for all manner of large initiatives during the 1960s, including all types of reforms. This consensus collapsed in the post-Vietnam and post-Watergate era of the 1970s, to a low of 26 percent of Americans believing that the government would seek to do right all or even most of the time by the end of the first decade of the 21st century.

There has been considerable research on the parts of society that embrace conspiracy theories of all types. Arguing that conspiracism writ large represents a fundamental part of the political system, legal scholar Mark Fenster claims that such conspiracies as the Moon hoax reflect an inadequacy that society as a whole is blind to, and that those who embrace these conspiracies are sublimating their frustrated ideals for the satisfaction of "knowing" something that is a secret from others. Fenster argues that at sum, denials of the Moon landings bring to the fore "a polarization so profound that people end up with an unshakable belief that those in power 'simply can't be trusted.'"

Various polls in the first decade of the new century all showed essentially the same thing: a tiny fraction of the population embraces these conspiracy ideas, and the overwhelming majority has always rejected them. In the United States, public opinion polls consistently have found that something in the neighborhood of 6 percent of the population doubt that the Moon landings occurred. It is important to note that "doubt" is not necessarily the same thing as "denial," but in any case, the small percentage is often in the realm of the margin for error. As one pollster said to me, "I can get 5 to 6 percent of the population to agree to anything." Gallup pollsters put it this way: "It is not unusual to find about that many people in the typical poll agreeing with almost any question that is asked of them; so the best interpretation is that this particular conspiracy theory is not widespread." A Gallup poll in 1999 found that "only 6% of the public believes the landing was faked and another 5% have no opinion." This tracked with other polls at other times that found a 6 percent denial rate.

In some societies, there has been a persistent effort to deny the Moon landings. In Cuba, for example, claims that the American Moon landings were faked are often promulgated. Cuba has been on the receiving end of a hard-edged US foreign policy that for more than 50 years tried to drive Fidel Castro from power. In such a context, believing the worst about Americans is an easy sell. As space journalist James Oberg has commented, "The results are similar, fanned by local attitudes toward the U.S. in general and technology in particular. Some religious fundamentalists—Hare Krishna cultists and some extreme Islamic mullahs, for example—declare the theological impossibility of human trips to other worlds in space."

This has expanded over time. In the summer of 2009, a British poll trumpeted the headline "Twenty-five per cent of the British public refuse to believe man has walked on the Moon, a survey conducted on behalf of E&T [Engineering & Technology] magazine has revealed." Personal experience supports this rising doubt. An email sent to me in August 2009 categorically declared the Moon landing to be a fake, adding, "The great majority of mankind is already familiar with the GREAT LIES & DECEPTIONS committed by these NASA HOAXERS about the 'Moon Landings.'" The writer commented, "I know for sure, that 'historians' are NOT the smartest people on this planet" before going on to deny the Holocaust "as an invention of HOLLYWOOD GANGSTERS . . . just like the 'Moon Landing.'" Perhaps historians are not always the smartest people on this planet, but such screeds do not demonstrate brilliance, even though this individual believes he has special knowledge that must be imparted to me and presumably to others.

During a fall 2009 trip to Vladivostok, Russia, I gave four presentations about the Moon landings to university students. In every case, the first question from the students was about challenges to the Moon-landing account made by conspiracy theorists. All of them channeled the standard criticisms of Apollo made by denialists over the years. Again, these might be more comprehensible in a setting where widespread distrust of the United States has been present for a century.

Perhaps this transition should not surprise those who study it. Many other truly weird beliefs exist in society. Gallup has reported that 37 percent of Americans believe that "houses can be haunted" and that 25 percent think that astrology can affect people's lives. In addition, 20 percent of Americans are not sure whether the Sun revolves around the Earth, and 49 percent believe the Earth is being visited by extraterrestrials. Those are statistics from scientific polls, but in nonrandom samples of questions about these subjects, the results get truly bizarre. A radio talk show host, Rob McConnell, declared that listeners to his show *The X Zone*, offered astounding responses to two questions—"Do you believe in ghosts, and did American astronauts really walk on the Moon?"—77 percent of respondents said yes to belief in ghosts, and 93 percent said that they did not believe that the Moon landings had actually occurred. As Seth Shostak from the SETI Institute remarked about this, "The respondents believe in ghosts, but do not think NASA put people on the moon. On the one hand, you have uncorroborated testimony about noises in the attic. On the other, you have a decade of effort by tens of thousands of engineers and scientists, endless rocket hardware, thousands of photos, and 378 kilograms (840 pounds) of moon rock." Shostak was befuddled by this reality of modern American society.

Apollo 17 astronaut Harrison Schmitt was more philosophical. "If people decide they're going to deny the facts of history and the facts of science and technology," he said, "there's not much you can do with them. For most of them, I just feel sorry that we failed in their education."

Folklorist Linda Degh asserted that the 1978 fictional feature film *Capricorn One*, in which NASA supposedly faked a landing on Mars, may have fostered greater acceptance of the denials of the Moon landings. No question, the February 2001 airing of the Fox television special *Conspiracy Theory: Did We Land on the Moon?* changed the nature of the debate. In this instance, a major network presented a conspiracy scenario without any serious rebuttal that might have been offered. As *USA Today* reported in the aftermath of the "news special":

According to Fox and its respectfully interviewed "experts"—a constellation of ludicrously marginal and utterly uncredentialed "investigative journalists"—the United States grew so eager to defeat the Soviets in the intensely competitive 1960s space race that it faked all six Apollo missions that purportedly landed on the moon. Instead of exploring the lunar surface, the American astronauts only tromped around a crude movie set that was created by the plotters in the legendary Area 51 of the Nevada desert.

While the program claimed to "let the viewer decide" about the validity of the claims for denial of the Moon landings, it made no attempt whatsoever to offer point and counterpoint, thereby giving the viewers a seriously biased view of the issue and skewed evidence in favor of a hoax.

The most egregious violation of propriety came in the Fox show when conspiracy theorists claimed that 10 astronauts and two civilians had died "under mysterious circumstances" during the Apollo program. The 10 astronauts in question included the three who were killed in the Apollo 1 fire of January 27, 1967, when their capsule was consumed in a flash fire during ground tests. At one point, Scott Grissom, son of Apollo 1 mission commander Gus Grissom, who was killed in the incident, claimed, "My father's death was no accident. He was murdered." At another point, however, he retracted that statement and declared that "NASA didn't murder anyone."

No question, shoddy workmanship and poor procedures led to those astronauts' deaths, and it was a tragic loss. But the deaths of Grissom, Ed White, and Roger Chaffee were an accident and not murder. Some Moon-landing deniers have claimed that NASA had them killed because Grissom was on the verge of exposing the Apollo program as a fraud. There is not one scintilla of evidence to support this accusation. The identity of the remaining seven astronauts presumably "murdered" by NASA during Apollo is something of a mystery. To be sure, astronaut Ed Givens died in an automobile accident, and astronauts Ted Freeman, C. C. Williams, Elliot See, and Charlie Bassett died in aircraft accidents, but these were far removed from the conduct of Project Apollo. This accounts for eight astronauts, but those making this claim also apparently count as astronauts two other pilots who had nothing to do with the Moon-landing program, X-15 pilot Mike Adams and Air Force Manned Orbiting Laboratory pilot Robert Lawrence. Finally, they claimed that NASA technician Thomas Baron and another NASA civilian, who was unnamed, were murdered to keep the secret of the Moon hoax. None of these claims was founded on any credible evidence.

The Fox television program fueled an explosion in conspiracy theories about the Moon landings and the audience for them. Glenn Johnson, who had an online model rocket business, remarked:

> The first I heard about it [the Moon-landing deniers] was a Fox TV special called something like "Did we ever go to the Moon?" which had all the feel of a legitimate documentary, but no real research was done. They talked to a bunch of people who were supposedly "experts" who had their reasons, unfounded as they were, for believing that the Apollo landings were impossible. Clever, really because no-one today has the ability to return to the sites to show what are there. So who are you going to believe? As in a court of law, all you have to show is "reasonable doubt" which is what they (claimed) to show.
>
> After that aired, it all kind of exploded on the internet and took on a life of its own. I'm sure that the people who had been making money talking about the "face on Mars" were jealous of the attention the Moon hoax groups were getting at that time. Being the resident "space geek," I was inundated with questions from friends and acquaintances—many of whom I thought were smarter than that—asking me to explain why the photos showed no stars and why the flag moved and the like. I did, but it got to be tiring, and unfortunately it changed my opinion that most people aren't so easily duped. They are.

Two scientists who have argued against the value of human space-flight even came forward to counter the Fox special's charges. Robert Park, director of the Washington office of the American Physical Society, dismissed the "documentary" with this statement: "The body of physical evidence that humans did walk on the Moon is simply overwhelming." Marc Norman at the University of Tasmania added, "Fox should stick to making cartoons. I'm a big fan of *The Simpsons!*"

The Fox television show exposed a much broader public than ever before to the arguments of the Moon-landing deniers. As folklorist Linda Degh noted, "The mass media catapult these half-truths into a kind of twilight zone where people can make their guesses sound as truths. Mass media have a terrible impact on people who lack guidance." Without a proper rebuttal available from NASA—the agency had an official position before of not responding to what it considered absurd claims—many young people publicly began to question the Apollo landings. Several astronauts stepped forward to affirm the legitimacy of the program, but others thought the charges too silly even to warrant response. Many debated the issues in the emerging world of the Internet. Indeed, the Internet became a haven for conspiracy theorists of all stripes, and with the barrier for publication online so low anyone could put up any page they wished with any assertions they wished to make. A search of the term "Moon hoax" recently yielded no fewer than 6,000 sites containing information of one type or another relating to this subject. But it also became a haven for counters to the conspiracy theorists, and a healthy debate has resulted.

Whereas NASA had refrained from officially responding to these charges—avoiding anything that might dignify the claims—the Fox show required that it change its approach. After the Fox program first aired, NASA released a one-paragraph press release entitled, "Apollo: Yes, We Did." It was minimalist, to say the least. It also posted a NASA information sheet originally issued in 1977 to readdress some of the concerns and pointed people with questions to various Internet sites where people responding to the deniers offered their rebuttals. Finally, NASA officials commented, "To some extent debating this subject is an insult to the thousands who worked for years to accomplish the most amazing feats of exploration in history. And it certainly is an insult to the memory of those who have given their lives for the exploration of space." This proved inadequate and the space agency soon created several Internet sites that addressed various aspects of the claims in the Fox television program.

As NASA chief historian at the time, I also took action, contracting a senior space writer, Jim Oberg, to write a detailed monograph responding to the Apollo hoax accusations as a primer for parents and teachers who wanted to answer questions from young people. I departed NASA in July 2002 for the Smithsonian Institution, but the project caught the media's attention that November. At the close of *ABC World News* on November 4, 2002, on the eve of national elections, Peter Jennings questioned that effort, saying that "NASA had been so rattled" that it commissioned "a book refuting the conspiracy theorists. . . . A professor of astronomy in California said he thought it was beneath NASA's dignity to give these Twinkies the time of day. Now, that was his phrase, by the way. We simply wonder about NASA." Jennings got this last part wrong. The professor was astrophysicist Phil Plait, and he supported NASA's making a response—he just wished it had not been necessary.

This commentary led to the cancellation of the monograph. NASA Administrator Sean O'Keefe did not want to deal with the negative media attention it received, despite the fact that educators and parents had been clamoring for such a primer. Oberg's comment on this situation is worth pondering:

> This is the way I see it: If many people who are exposed to the hoaxist arguments find them credible, it is neither the fault of the hoaxists nor of their believers—it's the fault of the educators and explainers (NASA among them) who were responsible for providing adequate knowledge and workable reasoning skills. And the localized success of the hoaxist arguments thus provides us with a detection system to identify just where these resources are inadequate.

In the summer of 2009, the United States celebrated the 40th anniversary of the Apollo 11 landing. NASA put together an aggressive celebration at various locations around the nation, especially in close proximity to NASA centers and in Washington, DC. This represented an excellent opportunity to assess the level of belief in the denials of the Moon landing among the population. Virtually every news story, especially in the electronic world, made some comment about a growing acceptance of denial of the landings. John Schwartz of the *New York Times* made light of this part of the coverage: "Forty years after men first touched the lifeless dirt of the Moon—and they did. Really. Honest.—polling consistently suggests that some 6 percent of Americans believe the landings were faked and could not have happened. The series of landings, one of the greatest gambles of the human race, was an elaborate hoax developed to raise national pride, many among them insist."

Others reported it more seriously. For example, the Associated Press issued a July 20, 2009, story that reported on the "small number" of

Moon-landing deniers. It quoted Bart Sibrel, who said, "It's 'an absolute fact' the astronauts didn't go to the moon." The story recounted Sibrel's efforts to sell videos on this subject, and remarked on the 2002 incident in which Buzz Aldrin punched him in the face for "accusing him of being a liar and a thief." The punch line: "Sibrel says he hopes President Barack Obama will confess and tell the world the truth." *Boston Globe* reporter Sam Allis waved in the direction of the deniers but then went on to offer a valentine to the success of the Moon landings:

> Apollo 11 is never dated. It transcended politics. There was a purity to the mission that secured its spot above the fray. It fused function and beauty. There was something very American about it, too. The mission was the product of our creativity, technical skills, and single-mindedness of purpose. A fractured country came together behind this effort. Name for me other efforts of such magnitude that have since brought us together. The human face of Apollo 11 was as much the guy in black frame glasses and a short-sleeve, no-iron, white shirt in Houston as it was Neil Armstrong. Apollo 11 was the triumph of the nerds and middle-class America. And it was romantic. A young, charismatic American president had challenged the country at the beginning of that decade to put a man on the moon by the end of the decade.

Allis recalled how his drill instructor in Army basic training camp had rousted him and his fellow draftees out of bed to watch the first steps on the Moon. Others throughout the anniversary also recalled their experiences of Apollo; of course, occasionally someone questioned the landings.

While not an empirical assessment my personal experience in dealing with numerous media inquiries, as well as public presentations concerning the Moon landings at the time of the various Apollo anniversaries suggests that the denials continue to resonate among the public. Virtually every journalist I talked with asked about the denials. Some made that the centerpiece of their articles. I tended to respond by asking the journalist why they asked about this, and the response was generally that they were looking for a new angle on the Apollo anniversary, and this gave them something different that had some punch to it. Accordingly, the flames of the Moon-landing denials were fanned by journalistic attention brought on by competition for a new and different perspective on the events.

The fact that the denials of the Moon landings would not go away should not surprise anyone. The twin features of modern society, a youth movement and postmodernism, helped to raise questions about the Moon landings. More than half the world's population was born after the last of the Moon landings took place in December 1972. Consequently, they had not lived through the excitement of the experience.

The media, especially, have fueled doubts over the years. While this may not be viewed as a definitive statement, a child's bib available at the time of the 45th anniversary of Apollo 11 read: "Once upon a time people walked on the moon. They picked up some rocks. They planted some flags. They drove a buggy around for a while. Then they came back. At least that's what grandpa said. The TV guy said it was all fake. Grandpa says the TV guy is an idiot. Someday, I want to go to the moon too." The indefatigable Oberg also commented on this phenomenon: "In the last ten years, an entirely new wave of hoax theories have appeared—on cable TV, on the Internet, via self-publishing, and through other 'alternative' publication methods. These methods are the result of technological progress that Apollo symbolized, now ironically fueling the arguments against one of the greatest technological achievements in human history."

CONCLUSION

Remembering Apollo

One of the hallmarks of recollections of the Apollo program among Americans has been proud nationalism and exceptionalism that often bordered on jingoism. That sense of nationalism is embarrassing in a postmodern, global age, and the elements of it that one still sees in recollections of Apollo suggests that Americans have yet to deal effectively with them. What were the ingredients of this unique form of nationalism, its origin as a Cold War response to Soviet space successes, and its role in promoting United States power to the rest of the world? Truly this nationalism involved an exclusionary and unique perspective on spaceflight and American technological virtuosity that found powerful expression in the nation's popular culture.

In a certain sense, our memory of Apollo is a myth, a "pattern of behavior" essentially held in common. This myth is not so much a fable or falsehood as it is a story, a kind of poetry, about events and situations that have great significance both for those involved and those that follow. Myths are, in fact, essential truths for the members of a cultural group who hold them, enact them, or perceive them. They are sometimes expressed in diffuse ideologies, but in literate societies such as the United States they are also embedded in historical narratives.

Memory, myth, and history are closely akin to each other; essentially they are stories that explain how things got to be the way they are. But common parlance suggests that memory is often faulty, myth is fiction, and only history is, or at least aspires to be, true. History to me, however, is an attempt to recount, model, or reconstruct the memory of the past for the purposes of the present. For a variety of reasons, such attempts are never completely successful. Thus, although few historians overtly do so, it is important to distinguish between history—the recounting of past events—and the past that is truly lost forever. History never fully or completely or accurately describes the past but instead attempts to develop approximate mental models or reconstructions of events. Different cultures at different times formulated and presented their reconstructions of the past in strikingly different ways. Thus, it is highly dangerous to attempt to evaluate the relationship between another culture's concept of "history," our own concept of history, and the lost reality of the past. All themes overlap in some way, but none is a precise mirror image of the other. Many people confuse history with the unrecoverable past and confuse myth and memory with fiction.

Alex Roland captured the importance of the Apollo myth best: It is a retelling for a specific purpose. It is not as much history as it is "tribal rituals, meant to comfort the old and indoctrinate the young." He noted:

All the exhilarating stories are here: the brave, visionary young President who set America on a course to the moon and immortality; the 400,000 workers across the nation who built the Apollo spacecraft; the swashbuckling astronauts who exuded the right stuff; the preliminary flights of Mercury and Gemini—from Alan Shepard's suborbital arc into space, through John Glenn's first tentative orbits, through the rendezvous and spacewalks of Gemini that rehearsed the techniques necessary for Apollo. There is the 1967 fire that killed three astronauts and charred ineradicably the Apollo record and the Apollo memory; the circumlunar flight of Christmas 1968 that introduced the world to Earth-rise over the lunar landscape; the climax of Apollo 11 and Neil Armstrong's heroic piloting and modest words, "that's one small step for a man, one giant leap for mankind"; the even greater drama of Apollo 13, rocked by an explosion on the way to the moon and converted to a lifeboat that returned its crew safely to Earth thanks to the true heroics of the engineers in Houston; and, finally, the anticlimax of the last Apollo missions.

Roland finds an epic aura of Apollo in this recitation of the voyages of discovery. The missions, however, turned into a dead end rather than a new beginning, and no amount of heroic prose could overcome that plot twist.

Central to any discussion of Apollo, mythic event or not, requires the consideration of the role of prestige in originating and sustaining the effort. Prestige, for all of its ubiquity in the literature of human spaceflight, is an imprecise term, and it perhaps obscures more than it illuminates. At sum, it signifies a demonstration of American superiority. But this superiority has many facets and audiences. It elicits both a "gut-level" reaction and calls for a more sophisticated explication. It is driven by politics of many sorts—international, bureaucratic, and domestic—none of them sufficient on their own to explain the primacy of human spaceflight in American culture, but all complexly intertwined.

Pride at home and prestige abroad were overwhelmingly significant rationales for undertaking an expansive Moon-landing program in the 1960s. In addition, there may well be four distinct attributes of the pride and prestige issue in Apollo. These include prestige on the international stage, using Apollo as a means for enhancing the attitudes of non-Americans toward the United States; pride at the national level, drawing the nation and its many peoples, priorities, and perspectives together; defining national identity, offering important ingredients into the national narrative celebrating exceptionalism among all else in the world; and embracing the idea of progress, and using the Apollo program as a symbol for American forward-thinking.

Almost from the beginning of thought about the potential of flight in space, theorists believed that the activity would garner worldwide prestige for those accomplishing it. For example, in 1946 the newly established RAND Corporation published the study *Preliminary Design of an Experimental World-Circling Spaceship*. This publication explored the viability of orbital satellites and outlined the technologies necessary for its success. Among its many observations, its comment on the prestige factor proved especially prescient: "A satellite vehicle with appropriate instrumentation can be expected to be one of the most potent scientific tools of the 20th Century. The achievement of a satellite craft would produce repercussions comparable to the explosion of the atomic bomb." In a paper published nine months later, RAND's James Lipp expanded on this idea: "Since mastery of the elements is a reliable index of material progress, the nation which first makes significant achievements in space travel will be acknowledged as the world leader in both military and scientific techniques. To visualize the impact on the world, one can imagine the consternation and admiration that would be felt here if the United States were to discover suddenly that some other nation had already put up a successful satellite."

This perspective is a classic application of what analysts have often referred to as "soft power." Coined by Harvard University professor Joseph Nye, the term gave a name to an alternative to threats and other forms of hard power in international relations. As Nye contends:

Soft power is the ability to get what you want by attracting and persuading others to adopt your goals. It differs from hard power, the ability to use the carrots and sticks of economic and military might to make others follow your will. Both hard and soft power are important . . . but attraction is much cheaper than coercion, and an asset that needs to be nourished.

In essence, activities such as Apollo represented a form of soft power, the ability to influence other nations through intangibles such as an impressive show of technological capability. It granted to the nation achieving it first, rightly as James Lipp forecast, an authenticity and gravitas not previously enjoyed among the world community. At root, this was an argument buttressing the role of spaceflight as a means of enhancing prestige on the world stage.

Even so, few appreciated the potential of spaceflight to enhance national prestige until the Sputnik crisis of 1957–58. Some have characterized that event as akin to Pearl Harbor, creating an illusion of a technological gap and providing the impetus for increased spending for aerospace endeavors, technical and scientific educational programs, and the chartering of new federal agencies to manage air and space research and development. This Cold War rivalry with the Soviet Union provided the key that opened the door to aggressive space exploration, not as an end in itself but as a means to achieving technological superiority in the eyes of the world. From the perspective of the 21st century it is difficult to appreciate the importance of the prestige factor in national thinking at the time.

John F. Kennedy responded to a perceived challenge of the Soviet Union by announcing the Apollo decision in 1961, and that rivalry sustained the effort. Kennedy put the world on notice that the United States would not take a back seat to its superpower rival. Kennedy said in 1962 that "we mean to be a part of it [spaceflight]—we mean to lead it. For the eyes of the world now look into space, to the moon and to the planets beyond, and we have vowed that we shall not see it governed by a hostile flag of conquest, but by a banner of freedom and peace. We have vowed that we shall not see space filled with weapons of mass destruction, but with instruments of knowledge and understanding." Apollo was a contest of wills, of political systems, of superpowers. And the United States intended to win it. Lyndon Johnson summed this up well with his assertion, "Failure to master space means being second best in every aspect, in the crucial area of our Cold War world. In the eyes of the world first in space means first, period; second in space is second in everything."

There is no question but that the Apollo program in particular, but also all of the human spaceflight efforts of the United States, was firstly about establishing US primacy in technology and with it the pride and prestige of great success. Apollo served as a surrogate for war, challenging the Soviet Union head on in a demonstration of technological virtuosity. The desire to win international support for the "American way" became the raison d'être for the Apollo program, and it served that purpose far better than anyone imagined when first envisioned. Apollo became first and foremost a Cold War initiative and aided in demonstrating the mastery of the United States before the world. This may be seen in a succession of Gallup polls conducted during the 1960s in which the question was asked: "Is the Soviet Union ahead of the U.S. in space?" Until the middle part of the decade, about the time that the Gemini program began to demonstrate American prowess in space, the answer was always that the United States trailed the Soviets. By the height of the Apollo Moon landings, world opinion had shifted overwhelmingly in favor of the United States. The importance of Apollo as an instrument of US foreign policy—which is not necessarily identical with national prestige and geopolitics but is closely allied—should not be mislaid in this discussion. It served, and continues to do so, as an instrument for projecting the image of a positive, open, dynamic American society abroad.

So for decades the United States launched humans into space for prestige, measured against similar Soviet accomplishments, rather than for practical scientific or research goals. This was in essence positive symbolism—each new space achievement acquired political capital for the United States, primarily on the international stage. As military analyst Caspar Weinberger noted in 1971, space achievements gave "the people of the world an equally needed look at American superiority."

At the same time Apollo heightened national pride. Observers have long recognized the impact that science and technology plays in building confidence in the US system in the international arena. Neither scholars nor politicians, however, have explicitly appreciated the importance of science and technology in building public confidence at home. Science and technology has been used to build public confidence in government in the same way that presidents use foreign policy to salvage their approval ratings. It can be argued that science and technology offer even greater opportunities for building public confidence than foreign adventures. It does so because scientific and technological programs tend to be driven by a clearly defined mission and a large degree of program autonomy that helps to close off much interest-group politics, factors that tend to favor program success.

Apollo, and perhaps spaceflight more generally, conjured images of the best in the human spirit and served, in the words of journalist Gregg Easterbrook, as "a metaphor of national inspiration: majestic, technologically advanced, produced at dear cost and entrusted with precious cargo, rising above the constraints of the earth." National pride served almost as much of a motivator of the US commitment to Apollo as the quest for global power. It may well be that space achievements, particularly those involving direct human presence, remain a potent source of national pride and that such pride is the underpinning reason why the American public continues to support human spaceflight and would find a decision to end the spaceflight program unacceptable. Certainly, space images—an astronaut on the Moon—rank just below the American flag and the bald eagle as patriotic symbols. The self-image of the United States as a successful nation is threatened when we fail in our space efforts, as we have seen from the embarrassment of a misshaped mirror on the Hubble Space Telescope to the sense of collective loss when some of our best citizens die before our eyes in space-shuttle accidents. Americans expect a successful program of human spaceflight as part of what the United States does as a nation. They are not overly concerned with the content or objectives of specific programs. But they are concerned that what is done seems worth doing and is done well. That sense of pride in space accomplishment has been missing in recent years.

It is somewhat trite to suggest that America was founded on the idea of progress and that it remains both an amorphous concept and one central to American national identity. In the 1830s an astute French interpreter of American society, Alexis de Tocqueville, observed that Americans had a "lively faith in human perfectibility." The manner in which these ideas have evolved over time has changed in relation to the larger society, and space exploration, especially Apollo, evinced these cherished conceptions.

Apollo had shown that virtually anything was possible. It suggested that America had both the capability and the wherewithal to accomplish truly astounding goals. All it needed was the will. As Senator Abraham Ribicoff mused in 1969, "If men can visit the Moon—and now we know they can—then there is no limit to what else we can do. Perhaps that is the real meaning of Apollo 11."

Space advocates also foresaw a new era of peace and mutual understanding arising in response to Apollo. Carl Sagan described "the unexpected final gift of Apollo" as "the inescapable recognition of the unity and fragility of the Earth." He added, "I'm struck again by the irony that spaceflight—conceived in the cauldron of nationalist rivalries and hatreds—brings with it a stunning transnational vision. You spend even a little time contemplating the Earth from orbit and the most deeply ingrained nationalisms begin to erode. They seem the squabbles of mites on a plum." Peaceful expansion beyond Earth, they thought, would lead to a golden age of peace and prosperity on Earth as national rivalries would be replaced with collective action toward a great positive goal. Forward-looking and aimed toward human perfections, Apollo signaled great opportunities for the future.

At its heart, this idea rested on what Richard T. Hughes has called the myth of the United States as a millennial nation, forever progressing to perfection. This sense of creating a perfect society has been present in American society from the very first. It also suggests that it is incumbent on those a part of this nation to further justice, equality, and liberty both inside and outside the confines of the United States. In many instances this is a positive set of attributes, as Hughes notes, but it might also be used to justify efforts "to export and impose its cultural and economics values throughout the world, regardless of the impact those policies might have on poor and dispossessed people in other parts of the world."

This sense of progress also embraced another national myth, that of the innocent nation. Without firm justification, the United States has come to believe that whatever it does is just and righteous and representative of best that humanity has to offer the world. This may be seen in virtually all periods of American history, but it is especially present in the great struggles of the 20th century. World Wars I and II especially led Americans to believe they were fighting for the survival of all that was good against forces of evil. But it also may be seen in the Cold War against the Soviet Union and in the aftermath of 9/11 in the global war on terrorism. So too, Apollo for most Americans is a representation of progress and eventual perfection.

There is a fine line between national pride and jingoism. Through the decade of the 1960s, national pride dominated much of the discussion of Apollo, even penetrating to popular culture. The late social commentator and comedian Sam Kinison once said to other nations seeking to undertake space spectaculars: "You really want to impress us! Bring back our Flag!" This is a particularly explicit expression of jingoism in relation to Apollo. While Kinison's challenge symbolized for many citizens US superiority and at the same time signaled the inferiority of all others, NASA situated its Apollo aspirations within the arena of international inclusiveness. If anything, these contradictory themes of national uniqueness and international cooperation have only intensified over time. This has long been critical to understanding the development of the space age.

Even without the crass jingoism of comedians, Apollo expresses those ideas quite effectively. Allusions to the "eagle" landing on the Moon, the placement of the US flag (and no other) on its surface, and the particular sense of possession Americans feel for the endeavor may be comfortable in the context of domestic affairs but are certainly increasingly uncomfortable as the space race recedes into history. Even so, the belief that humans would no longer be trapped on one tiny world, helpless to escape whatever might befall it, proved a powerful draw for those embracing spaceflight. As such, Apollo is celebrated as the first step off this world. Deeply mythic and powerfully nostalgic, Apollo's recollection in this manner resonates with the mythic frontier of the past that now morphs into the future frontier of space, a uniquely American perspective on this recent episode in history.

In part to advance the dominant narrative of Apollo as an American triumph, in part for mundane political reasons (such as expenditures in states where elections were up for grabs), in part to capture some of the aura of the Moon landings for a successive administration, and in part for a sense of the need to invest in science and technology because of the very real benefits that accrue through the process, several presidents and their political allies have advocated for a return to the Moon. These decisions deserve mention here as indications of the ongoing hold Apollo and the Moon has on the public policy of the United States.

Even before the end of the Apollo program in the early 1970s, some Americans were encouraging a return to the Moon to establish a lunar base. When Richard Nixon took office, he appointed the Space Task Group to study post-Apollo plans and make recommendations. Chartered on February 13, 1969, under the chairmanship of Vice President Spiro T. Agnew, this group met throughout the spring and summer to plot a course for the post-Apollo space program. The politics of this effort was intense. NASA lobbied hard with the group, and especially its chair, for a far-reaching post-Apollo space program that included development of a space station, a reusable Space Shuttle, a Moon base, and a human expedition to Mars. The NASA position was well reflected in the group's report on September 15, 1969, but Nixon did not act on the Space Task Group's recommendations. Instead, he was silent on the future of the US space program for several months and in the end did not accept the recommendation. Nixon issued a March 7, 1970, statement that clearly announced his approach toward dealing with NASA and space exploration: "We must also recognize that many critical problems here on this planet make high priority demands on our attention and our resources." Instead of continued lunar exploration, he approved a Space Shuttle only for Earth orbital operations.

This episode of post-Apollo planning demonstrates a negative impact of the Moon landings—an object lesson on how not to develop and sustain a long-term strategic program of human space exploration. It seems that in part because of the pressures to achieve Apollo on the schedule mandated by JFK, the technical solutions were not sustainable. At some level the program may also not have been sustainable because it was so exceptional. Even though pro-space advocates understood this problem, the political priority of Apollo thwarted the implementation of future expansive exploration efforts.

One might conceivably draw several conclusions from this episode. First, those leading NASA have a vision of what should be undertaken in human spaceflight that is always more expansive and less popular than what society as a whole, represented by the political leadership of the United States, is willing to undertake. Second, the technological challenge of bringing a new human space vehicle to reality is always greater than perceived at the time of the program's approval. Third, overpromising what might result from a particular program always has significant negative repercussions. That was certainly the story of any effort to return to the Moon, and we have seen it in numerous other settings as well.

The central question for NASA at the successful completion of Project Apollo was how best to continue its overall space exploration mission. This was a consideration especially important at the time because of the environment in Washington. As Apollo reached fruition, the nation was consumed with other crises: urban unrest, race riots, the Vietnam conflict and the antiwar movement, political radicalism of the left and the right, economic recession, welfare problems, and runaway budgets. Most important, the nation began to reassess the complex debate over the consensus of a vision of America as it had been articulated since the 1930s. This sustained criticism of national character and meaning plunged the United States into a fundamental divide over how to view the past, including Apollo. Political turmoil and activism led to the advance of a conservative perspective on politics, history, economics, and society, which emerged full-blown in the presidency of Ronald Reagan in the 1980s.

As this happened, Republicans in the White House—who served 26 of the years since the end of the Apollo program in 1972—embraced the legacy of Apollo and sought to replicate it on more than one occasion. Twenty years to the day after the Apollo 11 Moon landing, for example, Reagan's successor, President George H. W. Bush, made a Kennedy-like announcement by unveiling the ambitious Space Exploration Initiative (SEI) intended to return Americans to the Moon, establish a lunar base, and, then, using the space station and the Moon, reach Mars by the early part of the 21st century. The price tag for this effort was estimated at a whopping $400 billion over two decades, although some estimated the costs as high as $1 trillion. When briefed on this initiative, Stephen Kohashi, an aide to Senator Jake Garn (R-UT), exclaimed, "Have you lost your mind?" This demonstrated the constant push and pull between the desire of making measured, incremental progress in space exploration and aggressive funding increases that would both feed the NASA infrastructure and allow for aggressive Apollo-like programs.

Congress immediately reacted negatively. In votes for FY 1991 NASA funding, the SEI proposal was virtually zeroed out despite lobbying from Vice President Dan Quayle as the head of the National Aeronautics and Space Council, an advisory group to the president. Although the president castigated Congress for not "investing in America's future," members believed such a huge sum could be better spent elsewhere. Normally a strong supporter of NASA efforts, Maryland Senator Barbara Mikulski bluntly declared, "We're essentially not doing Moon–Mars."

In his support of SEI, Bush attempted to bring the space program full circle back to the early 1960s, but without the complementary elements that had made Kennedy's Apollo decision viable—that is, a crisis atmosphere that fostered the political will to do something spectacular, a favorable economic and technological climate, and tangible public support. In February 1993, looking for ways to cut the federal budget and thereby ease the federal deficit, the new Clinton administration announced that SEI did not fit into its plans for the space program. Accordingly, NASA Administrator Daniel S. Goldin said that the agency was not giving up on the idea of sending Americans back to the Moon and on to Mars, but simply "putting it off until we're ready and the nation is able to afford it." The fiscal year 1994 budget, therefore, contained no funding for SEI.

On January 14, 2004, President George W. Bush performed essentially a reenactment of his father by announcing a "Vision for Space Exploration" that called for humans to reach for the Moon and Mars during the next 30 years. As stated at the time, the fundamental goal of this vision was to advance US scientific, security, and economic interests through a robust space exploration program. In so doing, the president called for completion of an ongoing International Space Station (ISS) and retirement of the Space Shuttle fleet by 2010. Resources expended there would then go toward creating the enabling technologies necessary to return to the Moon and eventually to Mars. He also proposed a small increase in the NASA budget to help make this a reality. By 2008, however, it had become highly uncertain that the initiative could be realized, and in 2010 the program was cancelled by Bush's Democratic successor in the White House, Barack Obama. More recently, in December 2017 President Donald Trump directed NASA to "lead an innovative space exploration program to send American astronauts back to the moon, and eventually Mars." Trump added: "The directive I am signing today will refocus America's space program on human exploration and discovery. It marks an important step in returning American astronauts to the moon for the first time since 1972 for long-term exploration and use." He commented that this time a Moon program would lead to a permanent presence rather than just flags and footprints left behind. So far, however, there has been but a modest investment in making this objective a reality.

When President Trump made these statements, he was playing off a widely shared continuing desire to go to the Moon. The inertia and especially the cost soon lead to the realization, though, that this desire is not sufficiently strong to foster much in the way of action. Perhaps we have seen replays of the criticism that arose almost immediately with the Apollo decision in 1961. At that time, critics both on the political left and right questioned the huge investment that went into making the Moon landings a reality, despite whatever magnificence it might represent for the nation as a whole. With a price tag of $25.4 billion over an 11-year period, and more than $180 billion in 2017 dollars, it was the most expensive undertaking of the United States government in the 1960s except for national security. Opponents on both the left and right have asserted since the Kennedy administration that Apollo was inherently wasteful and should not be repeated. For example, William Proxmire attacked NASA's efforts to extend Apollo by undertaking space colonization, commenting, "It's the best argument yet for chopping NASA's funding to the bone. . . . I say not a penny for this nutty fantasy." Many conservatives likewise questioned the role of government in engaging in such exploits and believed that the money might be spent more productively elsewhere or given back to taxpayers as a reduction.

The euphoria over successfully completing the Moon landings has established such a powerful memory that most people in the United States reflecting on it believe that NASA enjoyed enthusiastic support during the 1960s and that somehow the agency lost its compass thereafter. Contrarily, at only one point before the Apollo 11 mission, October 1965, did more than half of the public favor the lunar landing program. Americans of the era consistently ranked spaceflight near the top of those programs to be cut in the federal budget. Such a position is reflected in public opinion polls taken in the 1960s when the majority of Americans ranked the space program as the government initiative most deserving of reduction, and its funding redistributed to Social Security, Medicare, and numerous other programs. While most Americans did not oppose space exploration per se, they certainly questioned spending on it when other problems appeared more pressing.

Since the heyday of Apollo, little has changed in this support for NASA and its human space exploration agenda. Many on the left view spaceflight, usually characterized as the human space program only, as a waste of resources that might be more effectively deployed to support other good ends. Many find themselves nodding in agreement when Josh Lyman, the White House assistant chief of staff in the fictional *West Wing* television series, told NASA officials in 2004 that his one priority for the space agency was that it stay out of the newspapers with tales of mismanagement and woe. He added that his agenda included using precious federal funds here on Earth to help people rather than to "conquer space."

Criticism of efforts to return to the Moon has taken myriad turns within the space community itself, as losers in the debate question the course taken. A subtext of this is that political decisions, especially Richard Nixon's decision to forego an expansive post-Apollo effort leading to a Moon base in the 1970s, lead directly to a conversation over the place of Apollo and a possible return to the Moon in the 21st century. As political scientist Howard E. McCurdy remarked, the Moon landing "was to America what the pyramids were to Egypt. It's one of our great accomplishments. . . . But when you go back and look, there were people, at the time who are expressing public misgivings. And in private—where you can get those kinds of conversations—[they] are pulling their hair out about this program."

The Moon landings were the result of a unique age in which the circumstances of politics, Cold War rivalries, the maturation of science and technology, a public willingness to embrace the endeavor, and the happenstance of an unfortunate presidential assassination converged to make possible the successful Apollo Moon landings. It seems unlikely that America will see another set of circumstances converge to create an environment conducive to a decision to send Americans to the Moon as a major national effort. There might be other ways in which humans might return there: perhaps a space race among other nations or perhaps a commercial rationale for a human expedition to return to the Moon, but otherwise it seems unlikely as we look toward the future of lunar exploration in the first part of the 21st century.

It is impossible to be unimpressed by the efforts made to reach the Moon in the 1960s and to revel in the success of humankind in achieving this striking accomplishment. A spring 1999 poll of opinion leaders sponsored by leading news organizations in the United States, for example, ranked the 100 most significant news events of the 20th century. The Moon landings came in a very close second to the splitting of the atom and its use during World War II. Some found the process of deciding between these various events difficult. "It was agonizing," CNN anchor and senior correspondent Judy Woodruff said of the selection process. The eminent historian and public intellectual Arthur M. Schlesinger Jr. summarized the position of many opinion leaders. "The one thing for which this century will be remembered 500 years from now was: This was the century when we began the exploration of space," he commented. Schlesinger said he looked forward toward a positive future and that prompted him to rank the lunar landing first. "I put DNA and penicillin and the computer and the microchip in the first 10 because they've transformed civilization. Wars vanish," he said. "Pearl Harbor will be as remote as the War of the Roses," he said, referring to the English civil war of the 15th century. He added, "The order is essentially very artificial and fictitious. It's very hard to decide the atomic bomb is more important than getting on the Moon."

In such a situation, Apollo increasingly seems to be something Americans did once upon a time for reasons that have receded far into the background. Returning to the Moon certainly will not happen again anytime soon without a major realignment of rationales, needs, and priorities. In 100 years, Apollo may be remembered as a singular event, glorious and revered but viewed increasingly as an undertaking without lasting significance. While advocates of human space exploration view the astronauts who landed on the Moon as people akin to 15th-century seafarers such as Christopher Columbus, the vanguards of sustained human exploration and migration, this opinion seems to diminish with every passing year. Early visions of human exploration were entertaining, but they were rooted in a relative lack of understanding about the nature of the Moon and planets and a view of colonization outmoded even at that time. The tradition of human space exploration may well continue, but probably not in the ways envisioned at the time of Apollo.

Some have characterized the Americans who fought and won World War II as "the greatest generation." Participants have been celebrated as heroes whether they carried a rifle in the Normandy invasion or built B-17 bombers in a factory. At the same time, no one would think that World War II was a great positive time. The death and destruction are something not to be repeated if possible. It would not surprise me to see recollections of Apollo echo that experience; participants will be venerated by those who were not present for the endeavor but who recognize it as a great achievement that will not be repeated. The trajectory may change, but probably not until there is a return to the Moon by humans and a discovery of something there of great worth.

ACKNOWLEDGMENTS

I wish to thank the many people—historians, other scholars and writers, archivists, and protagonists—for their assistance in many ways with the book. They include archivists at the NASA History Office who helped track down information and correct inconsistencies, as well as Steve Dick, Bill Barry, Steve Garber, and Nadine Andreassen at NASA; the staffs of the NASA Headquarters Library and the Scientific and Technical Information Program, who provided assistance in locating materials; Marilyn Graskowiak and her staff at the NASM Archives; and many archivists and scholars throughout NASA and other organizations. Many other archivists at the National Archives and Records Administration, especially at several of its Federal Records Centers and at the various presidential libraries, were also instrumental in making this study possible.

In addition to these individuals, I wish to acknowledge those who aided me in a variety of ways: Debbora Battaglia, Douglas Brinkley, Paul Ceruzzi, Angel Callahan, Erik Conway, Tom Crouch, Gen. John R. Dailey, Pete Daniel, David DeVorkin, Debbie Douglas, James Rodger Fleming, Alexander Geppert, G. Michael Green, James R. Hansen, Matt Hersch, Peter Jakab, Brian Jirout, Andy Johnston, Layne Karafantis, John Krige, W. Henry Lambright, Jennifer Levasseur, John Logsdon, Patrick McCray, Howard McCurdy, Emily A. Margolis, Teasel Muir-Harmony, Valerie Neal, Allan Needell, Michael Neufeld, Scott Pace, Robert Poole, Alan Stern, Harley Thronson, and Margaret Weitekamp. Assistance provided by a succession of interns helped to make this book a reality: Mary Bergen, Megan Bergeron, Meleta Buckstaff, Emily Gibson, Brian Jirout, Jessica Kirsch, Nicholas Limparis, Vicki Lindsey, Hermes Marticio, Maeve Montalvo, Ngoc Tran, and Heather van Werkhooven. My deep thanks are due to all of these fine people. All of them would disagree with some of the observations made here, but such is both the boon and the bane of historical inquiry.

I also thank those at Smithsonian Books for their work in bringing this book to fruition: Carolyn Gleason, Laura Harger, Matt Litts, Leah Enser, and Gregory McNamee. My greatest debts, of course, are to my daughters, Dana and Sarah Launius, and to my wife, Monique Laney.

NOTES

Prologue. July 20, 1969

xi: "I think you've got a fine-looking flying machine there": This and subsequent quotes about the landing are drawn from Edgar M. Cortright, ed., *Apollo Expeditions to the Moon* (Washington, DC: NASA SP-350, 1975), chap. 11.

xii: "My checklist says 1202": Ibid.

xii: "Altitude-velocity light": "Sounds from Apollo 11," www.nasa.gov/mission _pages/apollo/apollo11_audio.html (accessed May 30, 2018).

xiii: "That's one small step for [a] man": Cortright, *Apollo Expeditions to the Moon*, chap. 11.

xiv: "Here men from the planet Earth": "What Flag(s) on the Moon?" *New York Times*, June 23, 1969.

xv: "From space there is no hint of ruggedness": Michael Collins, *Carrying the Fire: An Astronaut's Journeys* (New York: Farrar, Straus and Giroux, 1974), 471.

xvi: "Before this decade is out": John F. Kennedy, "Urgent National Needs," *Congressional Record—House* (May 25, 1961), 8276.

xvii: "There is room for new cooperation": John F. Kennedy, final address to the United Nations General Assembly, www.americanrhetoric.com/speeches/ jfkunitednations1963.htm (accessed August 8, 2018).

xviii: "For better or for worse, your generation has been appointed": Lyndon B. Johnson, "Remarks at the University of Michigan," May 22, 1964, Public Papers of the Presidents, www.presidency.ucsb .edu/ws/?pid=26262%20 (accessed August 8, 2018).

xix: "Represent a significant new international achievement in space": George E. Mueller, NASA associate administrator for manned space flight, to Robert Gilruth, director, NASA Manned Spacecraft Center, November 4, 1968, NASA Manned Spacecraft Center Archives, Houston, TX.

xix: "Thanks, you saved 1968": Jeffrey Kluger, *Apollo 8: The Thrilling Story of the First Mission to the Moon* (New York: Henry Holt and Co., 2017), 288.

xx: "I kept racing between the TV and the balcony": "1969: 'One Small Step for Man,'" *On This Day*, news.bbc.co.uk/ onthisday/hi/witness/july/21/newsid _3058000/3058833.stm (accessed May 30, 2018).

1. Versions of Reality

2: "Somehow the problems which yesterday seemed large": Lyndon B. Johnson, "President's News Conference at the LBJ Ranch," August 29, 1965, *Public Papers of the Presidents: Lyndon B. Johnson, 1963–1969* (Washington, DC: United States Government Printing Office, 1969), 944–45.

2: "If we can put a man on the Moon, why can't we": Tom Horton, "On Environment: If America Could Send a Man to the Moon, Why Can't We . . . ?," *Baltimore Sun*, July 22, 1984.

3: "As it has been built up": Vannevar Bush to James E. Webb, administrator, NASA, April 11, 1963, 2, Presidential Papers, John F. Kennedy Presidential Library, Boston.

3: "Huge pile of resources": Amitai Etzioni, *The Moon-Doggle: Domestic and*

International Implications of the Space Race (New York: Doubleday, 1964), 70, 195.

4: "Why the great hurry to get to the moon": Dwight D. Eisenhower, "Are We Headed in the Wrong Direction?," *Saturday Evening Post*, August 11–18, 1962, 24.

4: "Has diverted a disproportionate share of our brain-power": Dwight D. Eisenhower, "Why I Am a Republican," *Saturday Evening Post*, April 11, 1964, 19.

5: "Products of the maniacal 1960s": Walter A. McDougall, "Technocracy and Statecraft in the Space Age," in *The Heavens and the Earth: A Political History of the Space Age* (New York: Basic Books, 1985), 1025.

5: "McDougall pictures [Eisenhower] as standing alone against the post-Sputnik stampede": Alex Roland, "How Sputnik Changed Us," *New York Times*, April 7, 1985, 1, 6.

5: "Perpetual technological revolution": McDougall, *The Heavens and the Earth*, 5–7.

2. A Moment in Time

8: "intellectual blood bank": Martin Weil and Emma Brown, "Ted Sorensen, JFK's Speechwriter and Defender, Dies at 82," *Washington Post*, November 1, 2010.

9: "I believe this nation should commit itself": Kennedy, "Urgent National Needs," 8276.

9: "The President looked strained": Theodore C. Sorensen, *Counselor: A Life at the Edge of History* (New York: Harper, 2008), 336.

10: "Routine applause": Theodore C. Sorensen, *Kennedy* (New York: Harper and Row, 1965), 526.

12: "Of all the major problems facing Kennedy": Hugh Sidey, *John F. Kennedy: Portrait of a President* (New York: Penguin, 1965), 98.

12: "Indeed, by having placed the highest national priority on the Mercury program": Jerome B. Wiesner, "Report to the President-Elect of the Ad Hoc Committee on Space," January 12, 1961, 16, Presidential Papers, John F. Kennedy Presidential Library, Boston.

13: "In the aftermath of that": T. Keith Glennan, *The Birth of NASA* (Washington, DC: NASA, 1993), 314–15.

13: "Do we have a chance of beating the Soviets": JFK, Memorandum for Vice President, April 20, 1961, Presidential Files, John F. Kennedy Presidential Library, Boston.

13: "If we can get to the moon before the Russians": *New York Times*, April 22, 1961.

15: "It's become a political struggle now": John F. Kennedy Presidential Library news release, "JFK Library Releases Recording of President Kennedy Discussing Race to the Moon," May 25, 2011, www.jfklibrary.org/About-Us/News-and-Press/Press-Releases/JFK-Library-Releases-Recording-of-President-Kennedy-Discussing-Race-to-the-Moon.aspx (accessed July 1, 2016).

17: "He was a hard man": Richard Reeves, "My Six Years with JFK," *American Heritage* 44, no. 7 (November 1993), www.americanheritage.com/content/my-six-years-jfk (accessed October 2, 2018).

19: "Join us in developing a weather prediction program": John F. Kennedy, "Annual Message to the Congress on the State of the Union, January 30, 1961," *American Presidency Project*, www.presidency.ucsb.edu/ws/index.php?pid=8045 (accessed October 15, 2018).

19: "We should offer the Soviets a range of choice": Eugene Skolnikoff, "President's Meeting with Khrushchev, Vienna, June 3–4, 1961, Reference Paper, Possible US-USSR Cooperative Projects," President's Office Files,

Countries: USSR, Vienna Meeting, Background Documents 1953–1961 (G-4), Briefing Material, Reference Papers, Box 126, John F. Kennedy Presidential Library, Boston.

19: "It is no secret that Kennedy would have preferred to cooperate": Aleksandr Fursenko and Timothy Naftali, *One Hell of a Gamble: The Secret History of the Cuban Missile Crisis* (New York: Norton, 1995), 121.

20: "I am not that interested in space": Tape recording of meeting between President John F. Kennedy and NASA Administrator James E. Webb, November 21, 1962, White House Meeting Tape 63, John F. Kennedy Presidential Library, Boston.

21: "We set sail on this new sea": President John F. Kennedy, "Address at Rice University on the Nation's Space Effort," September 12, 1962, Houston, John Fitzgerald Kennedy Presidential Library, Boston, www.cs.umb.edu/jfklibrary/j091262.htm (accessed May 12, 2017); italics original.

3. The Most Powerful Technology Ever Conceived

25: "Oh, the Apollo program": "Farouk El-Baz on the Apollo Program," National Academy of Engineering, www.engineeringchallenges.org/cms/8998/11830.aspx (accessed May 30, 2018).

25: "If we can put a man on the Moon": Horton, "On Environment."

26: "I believe that my generation has failed the American people": "Farouk El-Baz on the Apollo Program."

26: "Single-point failures in the sequence": Robert Gilruth Oral History no. 6 by David DeVorkin and John Mauer, March 2, 1987, Glennan-Webb-Seamans Project, National Air and Space Museum, Smithsonian Institution, Washington, DC.

33: "All-up": George E. Mueller, NASA, to Manned Spacecraft Center director et al., October 31, 1963; Eberhard Rees, Marshall Space Flight Center director, to Robert Sherrod, March 4, 1970, both in "Saturn 'All-Up' Testing Concept" File, Launch Vehicles, NASA Historical Reference Collection, NASA History Office, Washington, DC.

33: "Slipped more than six months": Maj. Gen. Samuel C. Phillips, Apollo program director, to J. Leland Atwood, president, North American Aviation, Inc., December 19, 1965, with attached "NASA Review Team Report," NASA Historical Reference Collection, NASA History Office, Washington, DC.

34: "Contributed a considerable amount of engineering innovation": George E. Mueller, "Joseph F. Shea," *Memorial Tributes: National Academy of Engineering* 10 (1999): 212.

34: "I have been in this business long enough": George E. Mueller, NASA associate administrator for manned space flight, to J. Leland Atwood, president, North American Aviation, Inc., December 19, 1965, NASA Historical Reference Collection, NASA History Office, Washington, DC.

35: "We have a fire in the cockpit": Sherl Larimer, "'We Have a Fire in the Cockpit!': The Apollo 1 Disaster 50 Years Later," *Washington Post*, January 26, 2017.

35: "From the foregoing": "Report of Apollo 204 Review Board," NASA Historical Reference Collection, NASA History Office, Washington, DC.

36: "We've always known that something like this was going to happen": Quoted in Erik Bergaust, *Murder on Pad 34* (New York: Putnam, 1968), 23.

37: "From this day forward, Flight Control will be known by two words": Quoted in Gene Kranz, *Failure Is Not an Option:*

Mission Control from Mercury to Apollo 13 and Beyond (New York: Berkeley Books. 2000), 204.

38: "And now they belong to the Heavens": Cartoon, *Washington Daily News,* January 28, 1967.

38: "I've been aboard on every flight": Cartoon, *Los Angeles Times,* January 31, 1967.

38: "In a Word—Carelessness": Cartoon, *Evening Star* (Washington, DC), April 11, 1967.

39: "Washington slept here": Cartoon, *Los Angeles Times,* April 12, 1967.

39: "It's official . . . you died from a case of lingering carelessness": Cartoon, *Washington Daily News,* April 12, 1967.

39: "Carelessness, not negligence, caused the Apollo fire": Cartoon, *Evening Bulletin* (Philadelphia), April 15, 1967.

39: "NASA 'would accept our part of the blame'": US Congress, House of Representatives, Subcommittee on NASA Oversight of the Committee on Science and Astronautics, "Investigation into Apollo 204 Accident," 90th Cong, 1st sess., April 10–12, 17, 21, May 10, 1967, 533.

39: "Evasiveness . . . lack of candor . . . patronizing attitude toward Congress": Senate Report 956, *Apollo 204 Accident: Report of the Committee on Aeronautical and Space Sciences, United States Senate, with Additional Views,* January 30, 1968, https://history.nasa.gov/as204_senate_956.pdf (accessed October 15, 2018).

39: "You are a menace and you are to blame for the fire": Quoted in Thomas Gordon White Jr., "The Establishment of Blame as Framework for Sensemaking in the Space Policy Subsystem: A Study of the Apollo 1 and Challenger Accidents," PhD diss., Virginia Polytechnic Institute and State University, 2000, 158.

40: "Joe Shea got up and started calmly": Christopher C. Kraft and James L.

Schefter, *Flight: My Life in Mission Control* (New York: Dutton, 2001), 275.

42: "A Gold Medal in the Lunar Olympics": Cartoon, *Seattle Times,* October 25, 1968.

42: "Almost Ready for the Ribbon-Cutting": Cartoon, *Huntsville* (AL) *Times,* October 23, 1968.

42: "Another Winner": Cartoon, *Los Angeles Herald-Examiner,* October 23, 1968.

42: "Splashdown on the Way Up": Cartoon, *Christian Science Monitor,* October 23, 1968.

43: "The command module was totally dominated": "Thomas J. Kelly Oral History," Johnson Space Center Oral History Collection, September 19, 2000, www.jsc.nasa.gov/history/oral_histories/KellyTJ/TJK_9-19-00.pdf (accessed February 21, 2018).

44: "The rest of us can only wait some desperate hour to hour": *Baltimore Sun,* April 15, 1970.

45: "Sanguine attitude about manned space flight": *Washington Post,* April 15, 1970, A1.

45: "Lindbergh gave up a continent": Editorial, *Washington Post,* April 15, 1970.

48: "In terms of numbers of dollars or of men": Dael Wolfe, "The Administration of NASA," *Science* 163 (November 15, 1968): 753.

50: "War on Untimely Death": Edward Everett Hazlett to Dwight D. Eisenhower, September 24, 1952, Eisenhower Personal Papers (1916–1953), Dwight D. Eisenhower Presidential Library, Abilene, KS.

50: "The old religious phenomenon of conversion": James B. Conant, "The Problems of Evaluation of Scientific Research and Development for Military Planning," speech to the National War College, February 1, 1952, quoted in James G. Hershberg, "'Over My

Dead Body': James B. Conant and the Hydrogen Bomb," paper presented to the Conference on Science, Military, and Technology, Harvard/MIT, June 1987, 50.

50: "If there was anything that bound the men": David Halberstam, *The Best and Brightest* (New York: Viking, 1973), 57, 153.

51: "Like any other group in our society, science has its full share of personalities": R. E. Lapp, *The New Priesthood: The Scientific Elite and the Uses of Power* (New York: Joanna Cotler Books, 1965), 227–28.

52: "A genuine facsimile of the Apollo 14 insignia": "In Memoriam: Carroll O'Connor and John Lee Hooker," *PBS News Hour*, June 22, 2001, www .pbs.org/newshour/bb/remember-jan -june01-oconnorhooker_06-22 (accessed August 8, 2018).

52: "They're talking about bioengineering animals and terraforming Mars": "The Sweet Smell of Air," *Sports Night*, first aired January 25, 2000.

4. Heroes in a Vacuum

53: "I still get a real hard-to-define feeling down inside when the flag goes by": John H. Glenn, "A New Era: May God Grant Us the Wisdom and Guidance to Use It Wisely," March 15, 1962, *Vital Speeches of the Day, 1962* (Washington, DC: US Government Printing Office, 1963), 324–26.

54: "[We] treated the men and their families with kid gloves": Dora Jane Hamblin to P. Michael Whye, January 18, 1977, NASA Historical Reference Collection, NASA History Office, Washington, DC.

55: "What made them so exciting": Quoted in Tom Wolfe, *The Right Stuff* (New York: Bantam Books, 1984), 94.

55: "The data on these early astronauts": Phyllis J. Johnson, "The Roles of NASA,

U.S. Astronauts and Their Families in Long-Duration Missions," *Acta Astronautica* 67 (2010): 561–71.

56: "Lead to a pilot making a wrong decision that might cost lives": Gordon Cooper and Bruce Henderson, *Leap of Faith: An Astronaut's Journey into the Unknown* (New York: HarperCollins, 2000), 26.

57: "If this were a military operation": James E. Webb to Robert R. Gilruth, April 15, 1965, James E. Webb Papers, Box 113, NASA—Astronaut Notes, Harry S. Truman Presidential Library, Independence, MO.

59: "David Scott, the Commander": Jeff Dugdale, "Moonwalkers," www .astrospacestampsociety.com/Articles01/ moonwalkers.html (accessed May 11, 2017).

60: "We were reprimanded and took our licks": David Scott and Alexei Leonov, *Two Sides of the Moon: Our Story of the Cold War Space Race* (New York: Pocket Books, 2005), 385–88.

60: "People used to tell me that I had no control over the astronauts": Robert Gilruth Oral History no. 6 by David DeVorkin and John Mauer, March 2, 1987, Glennan-Webb-Seamans Project, National Air and Space Museum.

61: "Wanted the entire agency to be faceless": Donn Eisele, *Apollo Pilot* (Lincoln: University of Nebraska Press, 2017), 69.

62: "Rumors surrounded Grissom": James Schefter, *The Race: The Complete True Story of How America Beat Russia to the Moon* (New York: Anchor, 2000), 72.

62: "The general trend": Kathy Keltner, "Mentions of Astronauts in Political Cartoons," September 13, 2002, unpublished study in possession of the author.

66: "Maybe we'll go next time": Cartoon, *Afro American*, December 17, 1972.

70: "They burned brightly in the glare of publicity": Marina Benjamin, *Rocket Dreams: How the Space Age Shaped Our Vision of a World Beyond* (New York: Free Press, 2003), 30–36.

71: "More of a miracle than a mechanical phenomenon": Norman Mailer, *Of a Fire on the Moon* (Boston: Little, Brown, 1970), 94–95.

71: "As a group, the public entertainments we tend to buy into": Center for Cultural Studies and Analysis, "American Perception of Space Exploration: A Cultural Analysis for Harmonic International and the National Aeronautics and Space Administration," presentation to NASA, April 21, 2004, Washington, DC, 23–24.

72: "The world does not, in fact, divide as neatly": Richard T. Hughes, *Myths America Lives By* (Champaign: University of Illinois Press, 2004), 186.

74: "The most perfect imaginable expression": Andrew Smith, *Moondust: In Search of the Men Who Fell to Earth* (New York: Fourth Estate, 2005), 295.

74: "The astronaut served": Susan Faludi, *Stiffed: The Betrayal of the American Man* (New York: HarperCollins, 1999), 452.

74: "NASA needed the pleasing faces, the frenzy of celebrity": Ibid., 451.

5. *Ex Luna, Scientia*

76: "Okay. Now let's go down and get that unusual one": "The Genesis Rock," *Apollo 15 Lunar Surface Journal*, www.hq .nasa.gov/office/pao/History/alsj/a15/ a15.spur.html (accessed February 22, 2018).

77: "Most of my thoughts on the Moon were of the geology involved": Quoted in Francis French, "An Eye on the Earth: Dave Scott and Pure Test Piloting," *Collectspace*, October 4, 2002, www .collectspace.com/news/news-100402a .html (accessed February 22, 2018).

77: "The direct scientific result of the Apollo Program": Paul D. Lowman Jr., "T Plus Twenty-Five Years: A Defense of the Apollo Program," *Journal of the British Interplanetary Society* 49 (1996): 76.

77: "A once-in-a-lifetime opportunity": Don E. Wilhelms, *To a Rocky Moon: A Geologist's History of Lunar Exploration* (Tucson: University of Arizona Press, 1993), 345–46.

78: "Was Apollo worth all the effort and expense": Smith, *Moondust*, 295–96.

78: "Many expressed disapproval of the manned program": Homer E. Newell, *Beyond the Atmosphere: Early Years of Space Science* (Washington, DC: US Government Printing Office, 1980), 209.

78: "We were manned if we did, and manned if we didn't": David H. DeVorkin, *Race to the Stratosphere: Manned Scientific Ballooning in America* (New York: Springer, 1989), 338.

80: "It is far cheaper and much easier": Willie Nash, "Mementos of Our 'Giant Leap,'" *Durham Herald-Sun*, July 20, 1999, A10.

81: "Moon rocks are absolutely unique": "The Great Moon Hoax," *Science@ NASA*, February 23, 2001, NASA Historical Reference Collection, NASA History Office, Washington, DC.

81: "The moon rocks were made in a NASA geology lab": Quoted in Rogier Van Bakel, "The Wrong Stuff," *Wired* (September 1994): 108–13, 155.

82: "From a scientific standpoint there seems little room for dissent": National Academy of Science Press Release NF6(100), "Man's Role in the National Space Program," August 7, 1961, NASA Historical Reference Collection, NASA History Office, Washington, DC.

82: "My position is that it is high time for a calm debate": James A. Van Allen, "Is Human Spaceflight Obsolete?" *Issues in Science and Technology* 20 (Summer 2004), issues.org/20-4/p_van_allen (accessed August 8, 2018).

83: "The lack of an adequate scientific endeavor": National Academy of Sciences/ National Research Council, *A Review of Space Research* (Washington, DC: NAS/ NRC Publication no. 1079, 1962), 11–13.

83: "The true purpose and fulfillment of life is to know and understand": G. Edward Pendray, *The Coming Age of Rocket Power* (New York: Harper Brothers, 1945), 226–27.

83: "The quest for scientific knowledge": Robert C. Seamans Jr., "The Challenge of Space Exploration," in *Smithsonian Treasury of 20th-Century Science,* ed. W. P. True (New York: Simon & Schuster, 1966), 50–51.

83: "To be preeminent in space": James E. Webb to President John F. Kennedy, November 30, 1962, NASA Historical Reference Collection, NASA History Office, Washington, DC.

84: "The most profound significance of Project Apollo is its catalytic effect": Ernst Stuhlinger, "Apollo: A Pattern for Problem Solving," in *Man on the Moon: The Impact on Science, Technology, and International Cooperation,* ed. Eugene Rabinowitch and Richard S. Lewis (New York: Basic Books, 1969), 195.

84: "We'll land, take a few photographs": Quoted in Donald A. Beattie, *Taking Science to the Moon* (Baltimore: Johns Hopkins University Press, 2001), xiv.

84: "Missed one of the very important elements necessary to the program": James E. Webb to Robert R. Gilruth, February 26, 1963, NASA Historical Reference Collection, NASA History Division, Washington, DC.

84: "Engineering tour-de-force": Quoted in R. Cargill Hall, "NASA: Thirty Years of Spaceflight," *Aerospace America* 26 (December 1988): 6–9.

85: "Where, for example, would astronauts be allowed to land": Robert Gillette and Allen L. Hammond, "Lunar Science: Letting Bygones Be Bygones," *Science* 179 (March 30, 1973): 1309.

86: "Seasonal changes there certainly were": V. A. Firsoff, *Strange World of the Moon* (New York: Basic Books, 1959), 172.

87: "How old is the Moon, how was it formed, and what is its composition": Beattie, *Taking Science to the Moon,* 13.

87: "In considering the lunar surface": TYCHO Study Group, "Report of August '65 'TYCHO' Meeting," 7–9, NASA contract no. NSR-24-005-047, November 23, 1965.

89: "The Moon moves through space as an ancient text": Harrison H. Schmitt, "What Have We Learned about the Moon," *Congressional Record—Extensions of Remarks,* February 24, 1975, E 651–53.

90: "Man's knowledge of the moon has been dramatically transformed": Allen L. Hammond, "Lunar Science: Analyzing the Apollo Legacy," *Science* 296 (March 30, 1973): 1313–15.

90: "The Moon is not a primordial object": Curator for Planetary Materials (no name supplied), Johnson Space Center, "Top Ten Scientific Discoveries Made during Apollo Exploration of the Moon," October 28, 1996, NASA Historical Reference Collection, NASA History Division, Washington, DC.

92: "According to this hypothesis": Stephen G. Brush, "A History of Modern Selenogony: Theoretical Origins of the Moon, from Capture to Crash, 1955–1984," *Space Science Reviews* 47 (1988): 214.

93: "Not because of any dramatic new development": D. J. Stevenson, "Origin of the Moon—The Collision Hypothesis,"

Annual Review of Earth and Planetary Sciences 15 (1987): 271–315.

93: "The giant-impact hypothesis appears to explain": Paul D. Spudis, "Moon," in *Encyclopedia of Space Science and Technology*, ed. Hans Mark (Hoboken, NJ: Wiley-Interscience, 2003), 2: 132.

93: "As it turned out, neither the Apollo astronauts": William K. Hartmann, R. J. Phillips, and G. J. Taylor, eds., *Origin of the Moon* (Houston: Lunar and Planetary Institute, 1986), vii.

93: "If our celebrated national lunar effort has produced any increased understanding": William K. Hartmann, "Where We Stand on the Moon," *Science* 169 (July 31, 1970): 465.

94: "Like so many others, I have stood in line at the Smithsonian Museum": Tim Challies, "Putting God in a Box—Recovering Awe," June 1, 2005, www.challies.com/articles/putting -god-in-a-box-recovering-awe (accessed May 11, 2017).

95: "We had people walking on the moon in 1969 through the 1970s": Randall H. Albright, "William Blake & Intellect, Reason, Science, Empiricism," July 8, 1996, www.albion.com/blake/archive/ volume1996/blake-d_Digest_V1996_83 .txt (accessed May 11, 2017).

95: "Than on any other night of the year": John Roach, "Full Moon Effect on Behavior Minimal, Studies Say," *National Geographic News Updated*, February 6, 2004, news.nationalgeographic.com/ news/2002/12/1218_021218_moon.html (accessed May 11, 2017).

95: "Speed the germination of your seeds": "Gardening by the Moon," www.gardeningbythemoon.com (accessed May 11, 2017).

96: "Meditation is a potent method of service to humanity": Alice A. Bailey, "Full Moon Meditation: An Introduction," www.ashtarcommandcrew.net/forum/ topics/full-moon-meditation-an -introduction-by-alice-a-bailey-this-full #ixzz4gnsyS8kL (accessed May 11, 2017).

97: "Are you any happier because men claimed they walked on the Moon": James A. Michener, *Space* (New York: Fawcett, 1982), 583.

6. Apollo Imagery and Vicarious Exploration

100: "They have done much, in the first place": Beaumont Newhall and Diana Edkins, *William H. Jackson* (Fort Worth, TX: Morgan & Morgan, 1974), 139.

101: "An index of national prestige and power": Stephen J. Pyne, "Seeking Newer Worlds: The Future of Exploration," 5, www.researchgate.net/ publication/265266038_SEEKING _NEWER_WORLDS_The_Future_of _Exploration (accessed May 11, 2017).

102: "It just seemed perfectly natural": "The Flight," *Time*, March 2, 1962.

103: "In determining the effectiveness": *Mercury Project Summary: Summary Including the Results of the Fourth Manned Orbital Flight, May 15 and 16, 1963* (Washington, DC: NASA SP-45, 1963), 220–21.

103: "We're looking at things that no human being had ever seen before": Richard W. Underwood Oral History, NASA Johnson Space Center Oral History Project, October 17, 2000, 5, www .jsc.nasa.gov/history/oral_histories/ UnderwoodRW/underwoodrw.pdf (accessed September 14, 2016).

104: "Selected cities, rail, highways, harbors, rivers, lakes": NASA Press Release 65-262, August 12, 1965; Gemini V file; NASA Historical Reference Collection, NASA History Division, Washington, DC. See also "Gemini 5 Crew Photographs Dallas, Tibet, Himalayas," *Aviation Week and Space Technology*, October 11, 1965, 31.

104: "You know, when you get back, you're going to be a national hero": Underwood Oral History, October 17, 2000, 19.

105: "The co-presence of two discontinuous elements": Roland Barthes, *Camera Lucida* (New York: Hill and Wang, 1981), 23.

105: "Photography, uniquely documentary and mass reproducible": Maren Stange, *Symbols of Ideal Life: Social Documentary Photography in America, 1890–1950* (Cambridge: Cambridge University Press, 1989), xiii.

107: "The reaches of space": Richard Nixon, "Inaugural Address, January 20, 1969," in *Public Papers of the Presidents of the United States: Richard Nixon—Containing the Public Messages, Speeches, and Statements of the President 1969* (Washington, DC: US Government Printing Office, 1971).

108: "Transcended countries and borders": "New Film about Neil Armstrong Omits American Flag from Moon Landing," *Fox News*, August 31, 2018, http://insider .foxnews.com/2018/08/31/first-man -new-film-about-neil-armstrong-omits -american-flag-moon-landing, accessed September 30, 2018.

108: "This is total lunacy": Rachel Leah, "Ryan Gosling Angers Marco Rubio with First Man American Flag Omission," *Salon*, August 31, 2018, https://www .salon.com/2018/08/31/ryan-gosling -angers-marco-rubio-with-first-man -american-flag-omission (accessed October 2, 2018).

108: "#proudtobeanamerican": Kim Willis, "Buzz Aldrin, second man on the moon, takes his shot in that 'First Man' American flag flap," *USA Today*, September 3, 2018, https://www .usatoday.com/story/life/movies/2018/ 09/03/first-man-buzz-aldrin-takes-his -shot-american-flag-flap/1185049002 (accessed October 2, 2018).

108: "The flag of the United States, and no other flag": "Implantation of the United States Flag on the Moon or Planets," P.L. 91-119, November 18, 1969, Section 8 (83 Stat., 202).

109: "Almost mystical unification of all people in the world at that moment": "Lunar Dust Smelled Just Like Gunpowder," *Life* 67 (August 22, 1969): 26.

110: "CowParade.com is your virtual 'moo-seum'": cowparade.com/AboutUs.php (accessed September 13, 2016).

111: "A photo of Aldrin, standing next to the crisp and bright flag": Richard Panchyk, "Flags," in *Americans at War: Society, Culture, and the Homefront, 1500–1815*, ed. John Phillips Resch (New York: Macmillan, 2004), 70.

112: "As the most evocative image from our landing on the Moon": "Armstrong and Tranquility Base Reflected in Visor," photograph 28, www.engology.com/ Apollo11.html (accessed May 11, 2017).

112: "First I took a picture of the surface of the Moon": "Solo Bootprint on the Moon," groups.google.com/forum/ #!topic/alt.sci.planetary/4kKmQXn-Lkw (accessed May 11, 2017).

115: "We had been in orbit for three orbits": NASA Interview with Bill Anders, Apollo 8, 1979, NASA Historical Reference Collection, NASA History Office, Washington, DC.

115: "We were the first humans to see the world in its majestic totality": Frank Borman and Robert J. Serling, *Countdown: An Autobiography* (New York: William Morrow/Silver Arrow Books, 1988), 204.

115: "Can there have been any more inspiring vision this century": James Lovelock, *Homage to Gaia: The Life of an Independent Scientist* (New York: Oxford University Press, 2001), 241.

116: "I turned on my blanket there on the gravel rooftop": Stewart Brand,

"The First Whole Earth Photograph," in
*Earth's Answer: Explorations of Planetary
Culture at the Lindisfarne Conferences*, ed.
Michael Katz, William Marsh, and Gail
Gordon Thompson (New York: Harper &
Row, 1977), 187.

118: "To see the Earth as it truly is": Oran
W. Nicks, ed., *This Island Earth*
(Washington, DC: NASA, 1970), 3.

7. Applying Knowledge from Apollo to This-World Problems

121: "While the decision to go to the moon
was unfolding": W. Henry Lambright,
Powering Apollo: James E. Webb of NASA
(Baltimore: Johns Hopkins University
Press, 1995), 99–100.

122: "I believe in the Democratic Party as a
vehicle for good government": James E.
Webb to Carl S. Day, February 10, 1961,
James E. Webb Papers, Harry S. Truman
Presidential Library, Independence, MO.

123: "One of the most important aspects of
the space program": James E. Webb,
NASA administrator, memorandum
for Program Officers, Headquarters,
and Directors, NASA Centers and
Installations, July 5, 1961, James
E. Webb Papers, Harry S. Truman
Presidential Library.

123: "The urgent necessity for a strong
technological underpinning": James
E. Webb, NASA administrator, to E.F.
Buryan, July 18, 1961, James E. Webb
Papers, Harry S. Truman Presidential
Library.

124: "Find ourselves on somewhat different
sides of the complex question": James E.
Webb to Vannevar Bush, May 15,
1961, NASA Historical Reference
Collection, NASA History Office,
Washington, DC.

125: "Would permit us to think of the country as having a complex in California":
James E. Webb, memorandum for the

vice president, May 23, 1961, NASA
Historical Reference Collection, NASA
History Office, Washington, DC.

125: "Grand mix of noble vision and pork-
barrel politics": Lambright, *Powering
Apollo*, 100.

125: "The best ways of utilizing the
tremendous developments of science":
James E. Webb, NASA administrator,
memorandum for Mr. Stoller, April 23,
1962, James E. Webb Papers, Harry S.
Truman Presidential Library.

125: "That the nation has, through its
democratic processes": James E. Webb,
NASA administrator, memorandum for
Mr. Drotning, July 23, 1962,
James E. Webb Papers, Harry S. Truman
Presidential Library.

126: "Eighty per cent of the professional and
technical personnel": James E. Webb,
NASA administrator, memorandum for
directors, NASA Field Centers, Western
Operations Office and North Eastern
Office, "The Industrial Applications
Program," September 19, 1962,
James E. Webb Papers, Harry S. Truman
Presidential Library.

127: "It seems to me, and I believe you
agree": James E. Webb, NASA administrator, to Lewis L. Strauss, May 12, 1961,
James E. Webb Papers, Harry S. Truman
Presidential Library.

127: "As a nation we are in a period of vast
and rapid change": James E. Webb,
NASA administrator, to Estes Kefauver,
January 14, 1963, James E. Webb Papers,
Harry S. Truman Presidential Library.

127: "Our society has reached a point where
its progress": James E. Webb, *Space Age
Management: The Large-Scale Approach*
(New York: McGraw-Hill, 1969), 15.

128: "142 universities have received grants":
NASA, *Summary Report: Future
Programs Task Group* (Washington, DC:
NASA, January 1965), 22.

128: "I became completely space crazy": Richard Monastersky, "Shooting for the Moon," *Nature* 460 (2009): 314–15.

129: "Little evidence was found that the Memorandums of Understanding": Office of Technology Utilization, Task Force to Assess NASA University Programs, *A Study of NASA University Programs* (Washington, DC: NASA Special Publication-185, 1968), 5–6.

130: "Science and technology have done to the city": NASA, *Conference on Space, Science, and Urban Life* (Washington, DC: NASA, 1963), 218.

130: "Multidisciplinary, large-scale effort": Webb, *Space Age Management*, 27.

131: "For more and better research in the area of management": Lt. Col. Robert H. Drumm, "Megamanagement for the Space Age," *Air University Review*, March–April 1970, www.airpower .maxwell.af.mil/airchronicles/aureview/ 1970/jul-aug/drumm.html (accessed March 16, 2008).

131: "Went a little far": In Piers Bizony, *The Man Who Ran the Moon: James E. Webb, NASA, and the Secret History of Project Apollo* (New York: Thunder's Mouth Press, 2006), 109–10.

131: "Have their report card marked against wobbly success standards": Thomas P. Paine, "Space Age Management and City Administration," *Public Administration Review* 29 (November–December 1969): 655.

132: "We do not oppose the Moon shot": *Titusville* (FL) *Star-Advocate*, July 15, 1969.

132: "We were coatless, standing under a cloudy sky": Thomas O. Paine, NASA administrator, Memorandum for Record, July 17, 1969, NASA Historical Reference Collection, NASA History Office, Washington, DC.

135: "Could be depended on to give a good account": William Hines, "Could NASA Solve All Problems?" *Birmingham* (AL) *News*, September 19, 1969.

136: "There is an intense power": "The Apollo Analogy," July 20, 2009, *Get Energy Smart! Now!*, getenergysmartnow .com/2009/07/20/the-apollo-analogy (accessed May 11, 2017).

136: "Go to the moon here on Earth in our push for technology": Dan Mauzy, "Kerry Ready to Fight in Divided Congress," 90.9 WBUR, Radio Boston, January 24, 2011, radioboston.wbur.org/2011/01/24/ john-kerry (accessed May 11, 2017).

8. Apollo and the Religion of Spaceflight

138: "An institutionalized collection of sacred beliefs": Robert Neelly Bellah, "Civil Religion in America," *Daedalus: Journal of the American Academy of Arts and Sciences* 96 (Winter 1967): 1–21.

139: "One sees elements of both a rational religion": Charles Reagan Wilson, "American Heavens: Apollo and the Civil Religion," *Journal of Church and State* 26, no. 2 (1984): 209–26, quote at 210.

139: "This step into the universe is a religion": Barbara Marx Hubbard, *The Hunger of Eve: One Woman's Odyssey toward the Future* (Harrisburg, PA: Stackpole Books, 1976), 144, 150.

139: "There is already too much religion in the space program": Memorandum from Charles E. Johnson to McGeorge Bundy, "Space Council's Reply to Questions of April 9, 1963," May 21, 1963, National Security Files, Box 307, John F. Kennedy Presidential Library, Boston.

139: "It is significant that a regime which preaches atheism above all else": H. A. van Helb, Dutch ambassador in Moscow, to Foreign Minister, the Hague, Netherlands, February 14, 1961, 2.05.248 inv 105, National Archives, the Hague, Netherlands.

140: "They don't know what to do when they get there": Mailer, *Of a Fire on the Moon*, 271.

141: "I have met an endangered species, and it is us": John W. Young, "The Big Picture: Ways to Mitigate or Prevent Very Bad Planet Earth Events," *Space Times: Magazine of the American Astronautical Society* 42 (November–December 2003): 22–23.

141: "Offer new places to live": Wernher von Braun, "For Space Buffs— National Space Institute, You Can Join," *Popular Science*, May 1976, 72–73.

142: "Spaceflight is first and foremost": Interview of Vadim Rygalov by Roger D. Launius, University of North Dakota, Grand Forks, April 24, 2008.

144: "The yin and yang of caution and boldness": Quoted in Steven J. Dick, "Risk and Exploration Revisited," NASA Headquarters, August 30, 2005, www .nasa.gov/exploration/whyweexplore/ Why_We_14.html (accessed October 23, 2012).

145: "A major center restructuring to accommodate Marshall's changing roles": "Dr. Rocco Petrone, Third Center Director, Jan. 26, 1973—March 15, 1974," history.msfc.nasa.gov/ management/center_directors/pages/ petrone1.html (accessed January 23, 2013).

145: "*Collier's* believes that the time has come": "What Are We Waiting For?" *Collier's*, March 22, 1952, 23.

147: "And that is how we do that": *Apollo 13*, film by Ron Howard, Universal Studios, 1995.

147: "Too many of us have lost the passion and emotion": Ray Bradbury, presenta- tion to annual meeting of the American Astronautical Society, Los Angeles, December 4, 1995.

148: "Yes, indeed, we are the lucky gener- ation": Walter Cronkite, "Our Infinite

Journey," in *Flight: A Celebration of 100 Years in Art and Literature*, ed. Anne Collins Goodyear, Roger D. Launius, Anthony M. Springer, and Bertram Ulrich (New York: Welcome Books, 2003), 231.

149: "Were coherent, mutually supportive, and reflective": Walter A. McDougall, *Promised Land, Crusader State: The American Encounter with the World since 1776* (Boston: Houghton Mifflin, 1997), 5.

149: "The old idea of American Christians as a chosen people": H. Reinhold Niebuhr, *The Kingdom of God in America* (1937; repr., New York: Harper and Row, 1959), 159.

9. Abandoned in Place

154: "A lot of people have tried over the years to save the tower": Todd Halvorsen, "NASA Delays Tower Destruction," *Florida Today*, February 7, 2004.

155: "The Apollo Mission Control Center is significant": Harry Butowsky et al., *Man in Space National Historic Landmark Theme Study* (Washington, DC: National Park Service, 2001), 57–58.

156: "Experiences gained by the Apollo astronauts": Ibid., 72–73.

157: "Langley's Lunar Landing Research Facility, completed in 1965": NASA LRC Fact Sheet, FS-1996-12-25-LaRC, "NASA Langley Research Center's Contributions to the Apollo Program," December 1996.

158: "This Agency has a dynamic research and development mission": G. Robert Nysmith, NASA associate administrator for management, to Ed Bearss, NPS chief historian, March 11, 1985, NASA Historical Reference Collection, NASA History Office, Washington, DC.

158: "NASA simply cannot afford to become entangled": G. Robert Nysmith, NASA associate administrator for management, to Raymond J. Nesbit, chair, NPS Advisory Board, April 19, 1985, NASA Historical Reference Collection, NASA History Office, Washington, DC.

159: "The question that the listing of technological facilities": Butowsky et al., *Man in Space National Historic Landmark Theme Study*, 53.

159: "Given the late-20th-century's pattern of rapid technological change": "Preserving Historic Scientific and Technological Facilities," *CRM Bulletin* 14, no. 2 (1991): 20–22.

161: "It would be an enormous folly": "Great Leap—to Museum?," *San Diego Union*, July 20, 1974.

161: "Although there are, at present, no plans for return visits to the moon": Walter L. Boyne, NASM director, to Russell Ritchie, NASA deputy associate administrator for external relations, October 25, 1984, Allan A. Needell Files, National Air and Space Museum, Smithsonian Institution, Washington, DC.

162: "There is also the bastard child of space business": Lou Dobbs with H. P. Newquist, *Space: The Next Business Frontier* (New York: iBooks, Inc., 2001), 9.

162: "Although this site is not yet 50 years in age": John Versluis, Jon Hunner, Ralph Gibson, and Beth O'Leary, New Mexico State University, to Robie Lange, National Historic Landmark Service, National Park Service, May 10, 1999, Allan A. Needell Files, National Air and Space Museum, Smithsonian Institution, Washington, DC.

163: "It has been determined as a matter of policy": Carol D. Shull, chief, National Historic Landmarks Survey, to John Versluis, New Mexico State University, June 8, 2000, and earlier drafts, in possession of the author.

164: "To establish special status for the Apollo 11 sites": Robert M. Kelso, Johnson Space Center, "NASA White Paper on the Preservation of U.S. Sites on the Lunar Surface," February 14, 2011, copy in possession of the author. This was the result of a workshop sponsored by NASA at the Kennedy Space Center, Florida, in January 2011, in which I participated as one of several individuals with specific ideas on how to create "rules of the road" for conduct of private-sector activities on the lunar surface.

164: "Approaching Apollo landing sites and artifacts at a tangent": Lucas Laursen, "NASA to Launch Guidelines to Protect Lunar Artifacts," *Science* 333 (September 2, 2011): 1207–08.

165: "Would be a far superior and long-lasting solution": "The Apollo Lunar Landing Legacy Act, H.R. 2617, introduced July 8, 2013; Henry R. Herzfeld and N. Scott Pace, 'International Cooperation on Human Lunar Heritage,'" *Science* 342 (November 29, 2013): 1049–50.

10. Denying the Apollo Moon Landings

167: "Doubted the moon voyage had taken place": Editorial, "Many Doubt Man's Landing on Moon," *Atlanta Constitution*, June 15, 1970.

167: "Three men floating inside a spaceship": Andrew Chaikin, *A Man on the Moon: The Voyages of the Apollo Astronauts* (New York: Viking, 1994), 100.

167: "27% expressed doubts that NASA went to the Moon": Mary Lynne Dittmar, "Building Constituencies for Project Constellation: Updates to the Market Study of the Space Exploration Program," presentation at "Building and Maintaining the Constituency for Long-Term Space Exploration," workshop,

George Mason University, Fairfax, VA,
July 31–August 3, 2006.
168: "The attribution of deliberate agency":
David Aaronovitch, *Voodoo Histories:
The Role of Conspiracy Theory in Shaping
Modern History* (New York: Penguin
Books, 2009), 5.
170: "It offended my sense of plausibility":
Ibid., 2.
171: "Just a month before, Apollo 11 astro-
nauts Buzz Aldrin and Neil Armstrong":
Bill Clinton, *My Life* (New York: Knopf,
2004), 244.
172: "A few stool-warmers in Chicago bars
are on record": John Noble Wilford, "A
Moon Landing? What Moon Landing?"
New York Times, December 18, 1969, 30.
172: "Many skeptics feel moon explorer
Neil Armstrong": "Many Doubt Man's
Landing on Moon."
172: "To some, the thrill of space can't
hold a candle": Howard A. McCurdy,
"Moonstruck," *Air & Space/Smithsonian*
(October–November 1998): 24.
173: "It's all a deliberate effort to mask problems at home": Quoted in Van Bakel,
"The Wrong Stuff," 155.
173: "You can't let one bit of information
out": Ibid., 112.
177: "Almost 40 years ago": "Top 10 Reasons
Why No Man Has Ever Set Foot on
the Moon," www.tarrdaniel.com/
documents/Ufology/moonhoax.html
(accessed May 12, 2017).
179: "About 36 lunar meteorites have
been found": Randy Korotev, "Lunar
Geochemistry as Told by Lunar
Meteorites," *Chemie der erde* 65 (2005):
297–346.
179: "The problem in recreating the Saturn
5": Michael Paine, "Saturn 5 Blueprints
Safely in Storage," March 13, 2000, www
.freerepublic.com/focus/news/1054183/
posts (accessed May 12, 2017).
180: "Neil Armstrong, the first man to
supposedly walk on the moon":

Quoted in "Bart Winfield Sibrel," www
.thekeyboard.org.uk/Bart%20Sibrel.htm
(accessed May 12, 2017).
180: "Their excuse: 'We thought it was true'":
Jesus Diaz, "Moronic Newspapers
Reprint *Onion*'s Neil Armstrong
Conspiracy Article as Fact," *Gizmodo*,
September 4, 2009, gizmodo.com/
5352887/moronic-newspapers-reprint
-onions-neil-armstrong-conspiracy
-article-as-fact (accessed May 12, 2017).
181: "A polarization so profound": Quoted in
John Schwartz, "Vocal Minority Insists
It Was All Smoke and Mirrors," *New
York Times*, July 13, 2009.
182: "Only 6% of the public believes the
landing was faked": Quoted in
Philip C. Plait, *Bad Astronomy:
Misconceptions and Misuses Revealed,
from Astrology to the Moon Landing Hoax*
(New York: John Wiley, 2002), 156.
182: "The results are similar": James Oberg,
"Lessons of the 'Fake Moon Flight'
Myth," *Skeptical Inquirer* (March/April
2003): 23, 30.
182: "Twenty-five per cent of the British
public refuse to believe": "Britons
Question Apollo 11 Moon Landings,
Survey Reveals," *Engineering &
Technology*, archive.is/0owYu (accessed
May 12, 2017).
183: "The respondents believe in ghosts":
Seth Shostak, "Public Split on Alien
Invaders and Spooky Specters," *Space.
com*, November 3, 2005, www.space.com/
1739-public-split-alien-invaders-spooky
-specters.html (accessed May 12, 2017).
183: "If people decide they're going to deny
the facts of history": Schwartz, "Vocal
Minority."
183: "According to Fox and its respectfully
interviewed 'experts'": Editorial, "Faking
a Hoax," *USA Today*, April 9, 2001.
185: "The first I heard about it":
RealSpaceModels to ProjectApollo
@yahoogroups.com, "Sorry to bring this

up, but I'm interested," July 24, 2006, copy in possession of the author.

185: "The body of physical evidence that humans": "The Great Moon Hoax," *Science@NASA*, February 23, 2001, solarviews.com/eng/moonhoax.htm (accessed May 12, 2017).

185: "The mass media catapult these half-truths": Quoted in Van Bakel, "The Wrong Stuff," 113.

186: "To some extent debating this subject is an insult": NASA Public Affairs, "RTQ Material/Talking Points 'Moon Hoax' Theories," February 15, 2001.

186: "NASA had been so rattled": Seth Borenstein, "Book to Confirm Moon Landings," *Deseret News* (Salt Lake City), November 1, 2002.

187: "This is the way I see it": Oberg, "Lessons."

187: "Forty years after men first touched the lifeless dirt of the Moon": Schwartz, "Vocal Minority."

188: "Apollo 11 is never dated": Sam Allis, "The King of All Trips: Apollo 11 Was a Lot of Good Things, Including Amazingly Cool," *Boston Globe*, July 19, 2009.

189: "In the last ten years, an entirely new wave of hoax theories": Oberg, "Lessons," 23.

Conclusion. Remembering Apollo

191: "All the exhilarating stories are here": Alex Roland, "How We Won the Moon," *New York Times Book Review*, July 17, 1994, 1, 25.

192: "A satellite vehicle with appropriate instrumentation": Project RAND, Douglas Aircraft Company, Engineering Division, *Preliminary Design of an Experimental World-Circling Spaceship* (SM-11827), May 2, 1946.

192: "Since mastery of the elements is a reliable index": Quoted in Virginia Campbell, "How RAND Invented the Postwar World: Satellites, Systems Analysis, Computing, the Internet—Almost All the Defining Features of the Information Age Were Shaped in Part at the RAND Corporation," *Invention & Technology* (Summer 2004): 53.

193: "Soft power is the ability to get what you want": Joseph S. Nye, "Propaganda Isn't the Way: Soft Power," *International Herald Tribune*, January 10, 2003.

193: "We mean to be a part of it": Kennedy, "Address at Rice University."

194: "Failure to master space means being second best": McDougall, *The Heavens and the Earth*, 1025.

194: "An equally needed look at American superiority": Caspar W. Weinberger to President Richard M. Nixon, via George Shultz, "Future of NASA," August 12, 1971, White House, Richard M. Nixon, president, 1968–1971 File, NASA Historical Reference Collection, NASA History Office, Washington, DC.

195: "A metaphor of national inspiration": Gregg Easterbrook, "The Space Shuttle Must Be Stopped," *Time*, February 2, 2003.

196: "Lively faith in human perfectibility": Alexis de Tocqueville, *Democracy in America*, ed. J. P. Mayer and Max Lerner (New York: Vintage Books, 1966), 343.

196: "If men can visit the Moon": *Congressional Quarterly* (July 25, 1969): 1311.

196: "The unexpected final gift of Apollo": Carl Sagan, *Pale Blue Dot: A Vision of the Human Future in Space* (New York: Random House, 1994), 171, 174–75.

196: "To export and impose its cultural and economics values": Richard T. Hughes, *Myths America Lives By* (Champaign: University of Illinois Press, 2004), 193.

198: "We must also recognize that many critical problems": *Public Papers of the President of the Unites States, Richard Nixon 1970* (Washington, DC:

US Government Printing Office, 1971), 250–53.

199: "Have you lost your mind?": Thor Hogan, *Mars Wars: The Rise and Fall of the Space Exploration Initiative* (Washington, DC: NASA SP-2007-4410, 2007), 66.

200: "Investing in America's future": "Bush Goes on the Counterattack against Mars Mission Critics," *Congressional Quarterly Weekly Report*, June 23, 1990, 1958.

200: "Putting it off until we're ready": "NASA Cancels Mars Trips," *Boston Globe*, April 10, 1993.

201: "Lead an innovative space exploration program": Dan Merica, "Trump to Send Astronauts Back to the Moon—and Eventually Mars," CNN, December 11, 2017, www.cnn.com/2017/12/11/politics/

trump-astronauts-moon/index.html (accessed October 2, 2018).

201: "It's the best argument yet for chopping NASA's funding": Robert Lovell, "Letters to L-5," *L-5 News* 2, no. 11 (November 1977): 1.

202: "Was to America what the pyramids were to Egypt": "Transcript: Washington Goes to the Moon, Part 1: 'Washington, We Have a Problem,'" aired May 25, 2001, WAMU-FM, transcript at wamu .org/d/programs/special/moon/opp _show.txt (accessed May 12, 2017).

203: "The one thing for which this century will be remembered": Arlene Levinson, "Atomic Bombing of Hiroshima Tops Journalists' List of Century's News," Associated Press, February 24, 1999, www.canoe.ca/CNEWSFeatures9902/ 24_news.html (accessed May 12, 2017).

ANNOTATED BIBLIOGRAPHY

Aldrin, Buzz. *Men from Earth*. New York: Bantam Books, 1989. Provides an intimate account of how NASA accomplished the national goal of putting an American on the Moon before the end of the decade. Aldrin's recounting of his two spaceflights is compelling, especially his telling of the nearly aborted Apollo 11 lunar landing.

———. *Return to Earth*. New York: Bantam Books, 1973. Describes the celebrity associated with being the second human on the Moon and Aldrin's sufferings with alcoholism and depression in the early 1970s. Aldrin talks about the pressure to keep the stress and day-to-day problems inside, and its effect on his marriage, which ended in divorce.

Allday, Jonathan. *Apollo in Perspective: Spaceflight Then & Now*. New York: Institute of Physics Publications, 1999. Takes a retrospective look at the Apollo space program and the technology that was used to land an American on the Moon as a means to explain the basic physics and technology of spaceflight.

Armstrong, Neil A., et al. *First on the Moon: A Voyage with Neil Armstrong, Michael Collins and Edwin E. Aldrin, Jr.* Written with Gene Farmer and Dora Jane Hamblin. Epilogue by Arthur C. Clarke. Boston: Little, Brown, 1970. This is the "official" memoir of the Apollo 11 landing mission to the Moon in 1969, and it contains much personal information about the astronauts that is not available elsewhere.

Bean, Alan L. *Apollo: An Eyewitness Account by Astronaut/Explorer/Artist/Moonwalker*. Shelton, CT: Greenwich Press Ltd., 1998. This is a large-format discussion of Apollo written by an Apollo 12 crewmember and illustrated with his unique artwork.

Beattie, Donald A. *Taking Science to the Moon: Lunar Experiments and the Apollo Program*. Baltimore: Johns Hopkins University Press, 2001. Gives a firsthand account of efforts by NASA scientists to do more to include science payloads on Apollo missions despite opposition from mission engineers, who envisioned a direct round-trip shot with as much margin for error as possible.

Benjamin, Marina. *Rocket Dreams: How the Space Age Shaped Our Vision of a World Beyond*. New York: Free Press, 2003. Ruminates on the scarred spaceflight culture that Apollo created and the later space program destroyed.

Benson, Charles D., and William Barnaby Faherty. *Moonport: A History of Apollo Launch Facilities and Operations*. Washington, DC: NASA SP-4204, 1978. Repr. as *Gateway to the Moon: Building the Kennedy Space Center Launch Complex* and *Moon Launch! A History of the Saturn-Apollo Launch Operations*, 2 vols. Gainesville: University Press of Florida, 2001. An excellent official history of the design and construction of the lunar launch facilities at Kennedy Space Center.

Bilstein, Roger E. *Stages to Saturn: A Technological History of the Apollo/Saturn Launch Vehicles*. Washington, DC: NASA SP-4206, 1980; rev. ed. 1996; repr., Gainesville: University Press of Florida, 2002. Based on exhaustive research and equipped with extensive bibliographic references, this book comes as close to being a definitive history of the Saturn rocket program as is likely ever to appear.

Borman, Frank, with Robert J. Serling. *Countdown: An Autobiography.* New York: William Morrow/Silver Arrow Books, 1988. Written to appear on the 20th anniversary of the first lunar landing, this autobiography covers much more than the Apollo program.

Brooks, Courtney G., James M. Grimwood, and Loyd S. Swenson Jr. *Chariots for Apollo: A History of Manned Lunar Spacecraft.* Washington, DC: NASA SP-4205, 1979. Based on exhaustive documentary and secondary research as well as 341 interviews, this well-written volume covers the design, development, testing, evaluation, and operational use of the Apollo spacecraft through July 1969.

Burgess, Colin. *Footprints in the Dust: The Epic Voyages of Apollo, 1969–1975.* Lincoln: University of Nebraska Press, 2010. Covers the flights of the Apollo program from Apollo 11 through the Apollo-Soyuz mission in 1975.

Burgess, Colin, and Kate Doolan, with Bert Viz. *Fallen Astronauts: Heroes Who Died Reaching for the Moon.* Lincoln: University of Nebraska Press, 2017 ed. Tells the stories of astronauts who died while employed by NASA.

Burrows, William E. *This New Ocean: The Story of the First Space Age.* New York: Random House, 1998. A comprehensive history of spaceflight.

Cadbury, Deborah. *Space Race: The Epic Battle between America and the Soviet Union for the Dominion of Space.* New York: HarperPerennial, 2007. A journalistic account of the race to the Moon.

CBS News. *10:56:20 PM EDT, 7/20/69: The Historic Conquest of the Moon as Reported to the American People.* New York: CBS, 1970. As its title suggests, this is an attempt to capture in print and pictures the reporting on humankind's first landing on the Moon during Apollo 11. More useful in capturing the immediacy of the moment than in providing an historical assessment of the event and its significance.

Cernan, Eugene, with Donald A. Davis. *The Last Man on the Moon: Astronaut Eugene Cernan and America's Race in Space.* New York: St. Martin's Press, 1999. A memoir that starts with Cernan's childhood days outside Chicago and continues through his college life at Purdue, his early career as a naval aviator, and his time as an astronaut.

Chaikin, Andrew. *A Man on the Moon: The Voyages of the Apollo Astronauts.* New York: Viking, 1994. One of the best books on Apollo, this work emphasizes the exploration of the Moon by the astronauts between 1968 and 1972.

Chapman, Richard L. *Project Management in NASA: The System and the Men.* Washington, DC: NASA SP-324, 1973. Based on almost 150 interviews and contributions by NASA officials, this volume provides a useful look at the agency's project management system, which contributed significantly to the success of the Apollo program.

Collins, Michael. *Carrying the Fire: An Astronaut's Journeys.* New York: Farrar, Straus and Giroux, 1974. The first candid book about life as an astronaut, written by the member of the Apollo 11 crew who remained in orbit around the Moon. Collins comments on other astronauts, describes the seemingly endless preparations for flights to the Moon, and assesses the results.

Compton, W. David. *Where No Man Has Gone Before: A History of Apollo Lunar Exploration Missions.* Washington, DC: NASA SP-4214, 1989. This clearly written account traces the ways in which scientists have researched the Moon.

Compton, W. David, and Charles D. Benson. *Living and Working in Space: A History of Skylab.* Washington, DC: NASA SP-4208,

1983. The official NASA history of Skylab, an orbital workshop placed in orbit in the early 1970s.

Cortright, Edgar M., ed. *Apollo Expeditions to the Moon*. Washington, DC: NASA SP-350, 1975. This large-format volume, with numerous color and black-and-white illustrations, contains essays by numerous luminaries, ranging from NASA Administrator James E. Webb to astronauts Michael Collins and Buzz Aldrin.

Ezell, Edward Clinton, and Linda Neuman Ezell. *The Partnership: A History of the Apollo-Soyuz Test Project*. Washington, DC: NASA SP-4209, 1978. An outstanding and detailed study of the effort by the United States and the Soviet Union in the mid-1970s to conduct joint human spaceflight.

Fowler, Eugene. *One Small Step: Project Apollo and the Legacy of the Space Age*. New York: Smithmark Publishers, 1999. A large-format coffee-table history that, rather than covering only the Apollo program itself, splits its contents almost evenly between the history of Apollo and the cultural impact of the space age.

Fries, Sylvia D. *NASA Engineers and the Age of Apollo*. Washington, DC: NASA SP-4104, 1992. A sociocultural analysis of a selection of the NASA engineers who worked on Project Apollo. The author makes extensive use of oral history in this study, providing both a significant appraisal of NASA during its golden age and important documentary material for future explorations.

Goldstein, Stanley H. *Reaching for the Stars: The Story of Astronaut Training and the Lunar Landing*. New York: Praeger, 1987. A detailed account of the development and management of the astronaut training program for Project Apollo.

Gray, Mike. *Angle of Attack: Harrison Storms and the Race to the Moon*. New York: Norton, 1992. This is a lively journalistic account of the career of Harrison Storms, president of the Aerospace Division of North American Aviation, which built the Apollo capsule. After the fire that killed three astronauts in January 1967, Storms was removed from the project. The book lays the blame at NASA's feet and argues that Storms and North American were scapegoats.

Hall, Eldon C. *Journey to the Moon: The History of the Apollo Guidance Computer*. Reston, VA: American Institute of Aeronautics and Astronautics, 1996. A detailed history of the development of the pioneering guidance computer built for the Apollo lunar module by MIT's Draper Laboratory, an effort in which the author was a senior participant.

Hallion, Richard P., and Tom D. Crouch, eds. *Apollo: Ten Years since Tranquillity Base*. Washington, DC: Smithsonian Institution Press, 1979. A collection of 16 essays published to commemorate the 10th anniversary of the first landing on the Moon on July 20, 1969; most of the contributions were written directly for the National Air and Space Museum by a variety of experts.

Hansen, James R. *First Man: The Life of Neil A. Armstrong*. New York: Simon & Schuster, 2005. The standard biography of Armstrong.

Hardesty, Von, and Gene Eisman. *Epic Rivalry: The Inside Story of the Soviet and American Space Race*. Washington, DC: National Geographic, 2007. A solid attempt to tell the story of the space race, written for the 50th anniversary of Sputnik.

Harford, James J. *Korolev: How One Man Masterminded the Soviet Drive to Beat America to the Moon*. New York: John Wiley & Sons, 1997. The first English-language biography of the Soviet "chief designer" who directed the projects that were so successful in the late 1950s and early 1960s in energizing the Cold War rivalry for space supremacy.

Harland, David M. *Exploring the Moon: The Apollo Expeditions.* Chichester, UK: Wiley-Praxis, 1999. This work focuses on the exploration and science missions carried out by Apollo astronauts while on the lunar surface.

Johnson, Stephen B. *The Secret of Apollo: Systems Management in American and European Space Programs.* Baltimore: Johns Hopkins University Press, 2002. This book skillfully interweaves technical details and fascinating personalities to tell the history of systems management in the United States and Europe. It is a very important work that uses Apollo as a key example.

Kauffman, James L. *Selling Outer Space: Kennedy, the Media, and Funding for Project Apollo, 1961–1963.* Tuscaloosa: University of Alabama Press, 1994. A straightforward history (but one that is quite helpful) of NASA's public image-building efforts and the relation of that image to public policy.

Kelly, Thomas J. *Moon Lander: How We Developed the Lunar Module.* Washington, DC: Smithsonian Institution Press, 2001. An outstanding memoir of the building of the lunar module, written by the Grumman engineer who led the effort.

Kluger, Jeffrey. *Apollo 8: The Thrilling Story of the First Mission to the Moon.* New York: Henry Holt and Co., 2017. A retelling of the Apollo 8 mission through the eyes of its crew.

Kraft, Christopher C., with James L. Schefter. *Flight: My Life in Mission Control.* New York: Dutton, 2001. Full of anecdotes, this memoir of Mission Control in Houston is most entertaining.

Kranz, Gene. *Failure Is Not an Option: Mission Control from Mercury to Apollo 13 and Beyond.* New York: Simon & Schuster, 2000. A good memoir of Mission Control.

Lambright, W. Henry. *Powering Apollo: James E. Webb of NASA.* Baltimore: Johns Hopkins University Press, 1995. This is an excellent biography of James E. Webb (1906–92), who served as NASA administrator between 1961 and 1968, the critical period during which Project Apollo was underway.

Launius, Roger D. *Apollo: A Retrospective Analysis.* Washington, DC: NASA SP-2004-4503, 1994; 2nd ed., 2004. A short study of Apollo's history, with key documents.

———. *NASA: A History of the U.S. Civil Space Program.* Anvil Series. Melbourne, FL: Krieger, 1994; rev. ed., 2001. This short history of US civilian space efforts consists of both narrative and documents and contains three chapters on the Apollo program.

Levine, Arnold S. *Managing NASA in the Apollo Era.* Washington, DC: NASA SP-4102, 1982. A narrative account of NASA from its origins through 1969, this book analyses the agency's key administrative decisions, contracting, personnel, the budgetary process, headquarters organization, relations with the Department of Defense, and long-range planning.

Liebergot, Sy, and David M. Harland. *Apollo EECOM: The Journey of a Lifetime.* Burlington, ON: Apogee Books, 2003. The autobiography of one of the key members of Mission Control in Houston during the Apollo program.

Light, Michael. *Full Moon.* New York: Knopf, 1999. Light weaves 129 stunningly clear images into a single composite voyage, a narrative of breathtaking immediacy and authenticity.

Lindsay, Hamish. *Tracking Apollo to the Moon.* New York: Springer Verlag, 2001. A history of the Apollo program from the perspective of an Australian involved in tracking the spacecraft that went to the Moon.

Logsdon, John M., ed. *Exploring the Unknown: Selected Documents in the History of the U.S. Civil Space Program.* 6 vols. Washington, DC: NASA Special Publication-4407, 1995–2004. An essential reference work reprinting more than 700 key documents in space policy and charting its development throughout the 20th century.

———. *John F. Kennedy and the Race to the Moon.* New York: Palgrave Macmillan, 2010. The definitive examination of Kennedy's role in sending Americans to the Moon.

Lovell, Jim, and Jeffrey Kluger. *Lost Moon: The Perilous Voyage of Apollo 13.* Boston: Houghton Mifflin, 1994. After the 1995 film *Apollo 13,* no astronaut had more fame than Jim Lovell, commander of the ill-fated mission to the Moon in 1970. This book is his recollection of the mission and the record on which the theatrical release was based.

Mackenzie, Dana, *The Big Splat, or How Our Moon Came to Be.* Hoboken, NJ: John Wiley, 2003. A fine discussion of how the science of Apollo led to a new interpretation of the Moon's origins.

Maher, Neil M. *Apollo in the Age of Aquarius.* Cambridge, MA: Harvard University Press, 2017. A major reinterpretation of the Apollo program and its relationship to the counterculture of the 1960s.

Mailer, Norman. *Of a Fire on the Moon.* Boston: Little, Brown, 1970. Mailer's book about the first lunar landing reflects his countercultural mindset and its antithesis, a NASA steeped in middle-class values and reverence for the American flag and culture.

Makemson, Harlen. *Media, NASA, and America's Quest for the Moon.* New York: Peter Lang, 2009. A study of media reporting on the lunar program.

McCurdy, Howard E. *Inside NASA: High-Technology and Organization Change in the U.S. Space Program.* Baltimore: Johns Hopkins University Press, 1993. A major study of changes in NASA's organizational culture from the Apollo era to the present.

———. *Space and the American Imagination.* Washington, DC: Smithsonian Institution Press, 1997. A pathbreaking study of the relationship between space and American culture.

McDougall, Walter A. *The Heavens and the Earth: A Political History of the Space Age.* New York: Basic Books, 1985. A Pulitzer Prize–winning book analyzing the race to the Moon in the 1960s. McDougall argues that Apollo prompted the space program to stress engineering over science, competition over cooperation, civilian over military management, and international prestige over practical applications.

Mindell, David. *Digital Apollo: Human and Machine in Spaceflight.* Cambridge, MA: MIT Press, 2011. An important study of the development of the Apollo guidance computer.

Mitchell, Edgar, and Ellen Mahoney. *Earthrise: My Adventures as an Apollo 14 Astronaut.* Chicago: Chicago Review Press, 2014. An astronaut memoir by an Apollo crewmember.

Mitchell, Edgar D., with Dwight Williams. *The Way of the Explorer: An Apollo Astronaut's Journey through the Material and Mystical Worlds.* New York: G. P. Putnam's Sons, 1996. Mitchell presents a smooth blend of memoir and exegesis, commenting at length on the ESP experiments he conducted on the flight and on his spiritual journey since returning to Earth.

Monchaux, Nicholas de. *Spacesuit: Fashioning Apollo.* Cambridge, MA: MIT Press, 2011. This scintillating and innovative book explores the layers of the spacesuit to tell the human story of its design and construction, as well as the stories of those who used it.

Montgomery, Scott L. *The Moon and the Western Imagination*. Tucson: University of Arizona Press, 1999. A richly detailed analysis of how the Moon has been visualized in Western culture through the ages, revealing the faces it has presented to philosophers, writers, artists, and scientists over nearly three millennia.

Murray, Charles A., and Catherine Bly Cox. *Apollo: The Race to the Moon*. New York: Simon & Schuster, 1989. Repr., Burkittsville, MD: South Mountain Books, 2004. Perhaps the best general account of the lunar program, this history uses interviews and documents to reconstruct the stories of the people who participated in Apollo.

Neufeld, Michael J. *Von Braun: Dreamer of Space, Engineer of War*. New York: Knopf, 2007. The standard work on the life of the rocket pioneer and the godfather of the Saturn V rocket, which took astronauts to the Moon.

Oberg, James. "Lessons of the 'Fake Moon Flight' Myth." *Skeptical Inquirer* (March/April 2003): 23, 30.

────. *Red Star in Orbit*. New York: Random House, 1981. Written by one of the premier Soviet space watchers, this history of the Soviet space program is among the best published in English prior to the fall of the Soviet Union in 1989, describing what was then known of the USSR's efforts to land a cosmonaut on the Moon before the Apollo landing in 1969.

Oliver, Kendrick. *To Touch the Face of God: The Sacred, the Profane, and the American Space Program, 1957–1975*. Baltimore: Johns Hopkins University Press, 2012. An underappreciated aspect of the ideology of human spaceflight and its relationship to religion.

Oreskes, Naomi, and John Krige, eds. *Science and Technology in the Global Cold War*.

Cambridge, MA: MIT Press, 2014. An important collection of essays, especially Asif A. Siddiqi's "Fighting Each Other: The N-1, Soviet Big Science, and the Cold War at Home."

Orloff, Richard G., ed. *Apollo by the Numbers: A Statistical Reference*. Washington, DC: NASA SP-2000-4029, 2000. An excellent statistical reference.

Paul, Richard, and Steven Moss. *We Could Not Fail: The First African Americans in the Space Program*. Austin: University of Texas Press, 2015. A major reinterpretation of the place of African American engineers and scientists in the Apollo program.

Pellegrino, Charles R., and Joshua Stoff. *Chariots for Apollo: The Making of the Lunar Module*. New York: Atheneum, 1985. A popular, though not always accurate, discussion of the development of the lunar module by Grumman Aerospace Corporation.

Poole, Robert. *Earthrise: How Man First Saw the Earth*. New Haven, CT: Yale University Press, 2008. A groundbreaking book on the Apollo 8 mission and the *Earthrise* photograph, which captured the global imagination.

Reynolds, David West. *Apollo: The Epic Journey to the Moon, 1963–1972*. New York: Zenith Press, 2013. Featuring a wealth of rare photographs, artwork, and cutaway illustrations, the book recaptures the excitement of the journey to the Moon.

Schirra, Wally, and Richard N. Billings. *Schirra's Space*. Annapolis, MD: Naval Institute Press, 1995. Wally Schirra was the only one of the original seven NASA astronauts to command a spacecraft in all three pioneering space programs—Mercury, Gemini, and Apollo.

Scott, David Meerman, and Richard Jurek. *Marketing the Moon: The Selling of the Apollo Lunar Program*. Cambridge, MA: MIT Press, 2014. An illustrated work on the sophisticated efforts by NASA and its many contractors to market the facts about space travel—through press releases, bylined articles, lavishly detailed background materials, and fully produced radio and television features—rather than push an agenda.

Shayler, David J. *Apollo: The Lost and Forgotten Missions.* Chichester, UK: Springer-Praxis, 2002. A discussion of planning for the aborted Apollo 18, 19, and 20 missions.

Shepard, Alan, and Deke Slayton. *Moonshot: The Inside Story of America's Apollo Moon Landings.* New York: Turner Publishing, 1994. Although it includes the recollections of two of the original Mercury Seven astronauts. chosen in 1959, this book is a disappointing as a general history of human space exploration by NASA from the first flight in 1961 through the last Apollo landing in 1972.

Siddiqi, Asif A. *Challenge to Apollo: The Soviet Union and the Space Race, 1945–1974.* Washington, DC: NASA SP-2000-4408, 2000; repr., 2 vols., Gainesville: University Press of Florida, 2003. The Soviet side of the race to the Moon.

———. *The Red Rockets' Glare: Spaceflight and the Soviet Imagination, 1857–1957.* New York: Cambridge University Press, 2010. A seminal study of the origins of the Soviet space program.

Slayton, Donald K., and Michael Cassutt. *Deke!: U.S. Manned Space, from Mercury to the Shuttle.* New York: Forge, 1995. The autobiography of one of the original Mercury Seven astronauts, selected in April 1959 to fly in space. Slayton served as a NASA astronaut during Projects Mercury, Gemini, Apollo, Skylab, and the Apollo-Soyuz Test Project, but his only spaceflight took place in July 1975.

Smith, Andrew. *Moondust: In Search of the Men Who Fell to Earth.* New York: Fourth Estate, 2005. The author interviewed all the remaining Apollo astronauts, seeking to learn how their lives had changed because of the experience.

Stafford, Thomas P., and Michael Cassutt. *We Have Capture: Tom Stafford and the Space Race.* Washington, DC: Smithsonian Institution Press, 2002. Stafford flew four times—on Gemini VI and IX, Apollo 10, and the Apollo-Soyuz Test Project—but was especially significant for his efforts, beginning in the 1970s, as the unofficial ambassador to the Soviet Union for space and his key roles in defining space policy in the United States.

Steven-Boniecki, Dwight. *Live TV from the Moon.* Burlington, ON: Apogee, 2010. Covers the earliest known proposals for TV coverage on lunar missions and the constant battles over the inclusion of the TV system on Apollo missions.

Sullivan, Scott P. *Virtual Apollo: A Pictorial Essay of the Engineering and Construction of the Apollo Command and Service Modules.* Burlington, ON: Apogee Books, 2003. A collection of exceptionally accurate drawings of Apollo hardware.

Swanson, Glen E., ed. *"Before This Decade Is Out . . .": Reflections on the Apollo Program.* Washington, DC: NASA Special Publication-4223, 1999; repr., Gainesville: University Press of Florida, 2002. A collection of oral histories with some of the key individuals associated with Project Apollo, including George Mueller, Gene Kranz, James Webb, and Wernher von Braun.

Tribbe, Matthew D. *No Requiem for the Space Age: The Apollo Moon Landings and American Culture.* New York: Oxford University Press, 2014. Offers a portrait of a nation questioning its values and capabilities, with Apollo as the center of the debate.

Turnill, Reginald. *The Moonlandings: An Eyewitness Account.* New York: Cambridge University Press, 2002. Longtime BBC aerospace reporter Turnill gives a comprehensive overview of the Apollo program, including its origins in America's post-Sputnik panic, as well as the program's demise amid waning public interest, rising costs, and a general sense that the Moon launches had accomplished all they could.

Webb, James E. *Space Age Management: The Large-Scale Approach.* New York: McGraw-Hill, 1969. Based on a series of lectures, this book by the former NASA administrator tries to apply the concepts of large-scale technological management employed in Apollo to society's other problems.

Wendt, Guenter, and Russell Still. *The Unbroken Chain.* Burlington, ON: Apogee Books, 2001. Memoirs are in vogue for the Apollo pioneers. Guenter Wendt was NASA's legendary "pad leader" for all human space launches from the first Mercury mission in 1961 through the last Apollo flights.

Westwood, Lisa, Beth Laura O'Leary, and Milford Wayne Donaldson. *The Final Mission: Preserving NASA's Apollo Sites.* Gainesville: University Press of Florida, 2017. A discussion of the historic sites of the Apollo program and how they might be preserved.

White, Thomas Gordon Jr. "The Establishment of Blame as Framework for Sensemaking in the Space Policy Subsystem: A Study of the Apollo 1 and Challenger Accidents." Ph.D. dissertation, Virginia Polytechnic Institute and State University, 2000. A strong assessment of the manner in which organizations make sense of tragedies and assign blame to entities.

Wilford, John Noble. *We Reach the Moon: The New York Times Story of Man's Greatest Adventure.* New York: Bantam Books, 1969. One of the earliest journalistic accounts to appear at the time of Apollo 11, a key feature of this general and workmanlike (but undistinguished) history is a 64-page color insert of mission photos. It was prepared by a *New York Times* science writer using his past articles.

Wilhelms, Don E. *To a Rocky Moon: A Geologist's History of Lunar Exploration.* Tucson: University of Arizona Press, 1993. This detailed account of lunar exploration and science provides a detailed and contextual account of lunar geology during the 1960s and 1970s and an engaging story of the scientific exploration of the Moon as seen by one of the field's more important behind-the-scenes scientists.

Worden, Al, and Francis French. *Falling to Earth: An Apollo 15 Astronaut's Journey to the Moon.* New York: Smithsonian Books, 2011. As command module pilot for the Apollo 15 mission to the Moon in 1971, Worden flew on what is widely regarded as the greatest exploration mission that humans have ever attempted.

Young, John W., with James R. Hansen. *Forever Young: A Life of Adventure in Air and Space.* Gainesville: University Press of Florida, 2012. An astronaut's experiences in Gemini, Apollo, and beyond.

Zimmerman, Robert. *Genesis: The Story of Apollo 8.* New York: Four Walls Eight Windows, 1998. A detailed account of the December 1968 circumlunar mission by Frank Borman, Bill Anders, and Jim Lovell.

INDEX

Page numbers in italics refer to photographs.